Seeing and Touching Structural Concepts

Seeing and Touching Structural Concepts

Tianjian Ji and Adrian Bell

Taylor & Francis
Taylor & Francis Group
LONDON AND NEW YORK

First published 2008
by Taylor & Francis
2 Park Square, Milton Park, Abingdon, Oxon OX14 4RN

Simultaneously published in the USA and Canada
by Taylor & Francis
270 Madison Ave, New York, NY 10016

Taylor & Francis is an imprint of the Taylor & Francis Group, an informa business

© 2008 Tianjian Ji and Adrian Bell

Typeset in Sabon by Wearset Ltd, Boldon, Tyne and Wear
Printed and bound in Great Britain by TJ International Ltd, Padstow, Cornwall

All rights reserved. No part of this book may be reprinted or reproduced or utilised in any form or by any electronic, mechanical, or other means, now known or hereafter invented, including photocopying and recording, or in any information storage or retrieval system, without permission in writing from the publishers.

The publisher makes no representation, express or implied, with regard to the accuracy of the information contained in this book and cannot accept any legal responsibility or liability for any efforts or omissions that may be made.

British Library Cataloguing in Publication Data
A catalogue record for this book is available from the British Library

Library of Congress Cataloging-in-Publication Data
Ji, Tianjian.
Seeing and touching structural concepts/Tianjian Ji and Adrian Bell.
p. cm.
Includes bibliographical references and index.
1. Structural analysis (Engineering) I. Bell, Adrian. II. Title.
TA645.J53 2008
624.1′71–dc22   2007044947

ISBN10: 0-415-39773-1 (hbk)
ISBN10: 0-415-39774-X (pbk)
ISBN10: 0-203-96079-3 (ebk)

ISBN13: 978-0-415-39773-5 (hbk)
ISBN13: 978-0-415-39774-2 (pbk)
ISBN13: 978-0-203-96079-0 (ebk)

Contents

Preface xii
Acknowledgements xvi

PART I
Statics 1

1 Equilibrium 3

   1.1 *Definitions and concepts 3*
   1.2 *Theoretical background 3*
   1.3 *Model demonstrations 5*
       1.3.1 *Action and reaction forces 5*
       1.3.2 *Stable and unstable equilibrium 6*
       1.3.3 *A plate–bottle system 7*
       1.3.4 *A magnetic 'float' model 7*
   1.4 *Practical examples 8*
       1.4.1 *A barrier 8*
       1.4.2 *A footbridge 9*
       1.4.3 *An equilibrium kitchen scale 10*
       1.4.4 *Stage performance 10*
       1.4.5 *Magnetic float train 11*
       1.4.6 *A dust tray 11*

2 Centre of mass 13

   2.1 *Definitions and concepts 13*
   2.2 *Theoretical background 13*
   2.3 *Model demonstration 18*
       2.3.1 *Centre of mass of a piece of cardboard of arbitrary shape 18*
       2.3.2 *Centre of mass and centroid of a body 19*
       2.3.3 *Centre of mass of a body in a horizontal plane 19*
       2.3.4 *Centre of mass of a body in a vertical plane 21*
       2.3.5 *Centre of mass and stability 22*
       2.3.6 *Centre of mass and motion 24*

2.4 Practical examples 24
  2.4.1 Cranes on construction sites 24
  2.4.2 The Eiffel Tower 25
  2.4.3 A display unit 26
  2.4.4 The Kio Towers 26

3 Effect of different cross-sections 28

3.1 Definitions and concepts 28
3.2 Theoretical background 28
3.3 Model demonstrations 33
  3.3.1 Two rectangular beams and an I-section beam 33
  3.3.2 Lifting a book using a bookmark 34
3.4 Practical examples 35
  3.4.1 A steel-framed building 35
  3.4.2 A railway bridge 36
  3.4.3 I-section members with holes (cellular beams and columns) 36

4 Bending 38

4.1 Definitions and concepts 38
4.2 Theoretical background 38
4.3 Model demonstration 42
  4.3.1 Assumptions in beam bending 42
4.4 Practical examples 43
  4.4.1 Profiles of girders 43
  4.4.2 Reducing bending moments using overhangs 43
  4.4.3 Failure due to bending 44
  4.4.4 Deformation of a staple due to bending 45

5 Shear and torsion 47

5.1 Definitions and concepts 47
5.2 Theoretical background 47
  5.2.1 Shear stresses due to bending 47
  5.2.2 Shear stresses due to torsion 49
5.3 Model demonstrations 53
  5.3.1 Effect of torsion 53
  5.3.2 Effect of shear stress 53
  5.3.3 Effect of shear force 55
  5.3.4 Open and closed sections subject to torsion with warping 56
  5.3.5 Open and closed sections subject to torsion without warping 57
5.4 Practical examples 58
  5.4.1 Composite section of a beam 58
  5.4.2 Shear walls in a building 58
  5.4.3 Opening a drinks bottle 59

6 Stress distribution — 61

    6.1 Concept 61
    6.2 Theoretical background 61
    6.3 Model demonstrations 62
        6.3.1 Balloons on nails 62
        6.3.2 Uniform and non-uniform stress distributions 63
    6.4 Practical examples 64
        6.4.1 Flat shoes vs. high-heel shoes 64
        6.4.2 The Leaning Tower of Pisa 65

7 Span and deflection — 67

    7.1 Concepts 67
    7.2 Theoretical background 67
    7.3 Model demonstrations 71
        7.3.1 Effect of spans 71
        7.3.2 Effect of boundary conditions 72
        7.3.3 The bending moment at one fixed end of a beam 73
    7.4 Practical examples 74
        7.4.1 Column supports 74
        7.4.2 The phenomenon of prop roots 75
        7.4.3 Metal props used in structures 75

8 Direct force paths — 77

    8.1 Definitions, concepts and criteria 77
    8.2 Theoretical background 77
        8.2.1 Introduction 77
        8.2.2 Concepts for achieving a stiffer structure 78
        8.2.3 Implementation 82
        8.2.4 Discussion 87
    8.3 Model demonstrations 90
        8.3.1 Experimental verification 90
        8.3.2 Direct and zigzag force paths 92
    8.4 Practical examples 92
        8.4.1 Bracing systems of tall buildings 92
        8.4.2 Bracing systems of scaffolding structures 93

9 Smaller internal forces — 96

    9.1 Concepts 96
    9.2 Theoretical background 96
        9.2.1 Introduction 96
        9.2.2 A ring and a tied ring 97
    9.3 Model demonstrations 105

9.3.1 A pair of rubber rings 105
9.3.2 Post-tensioned plastic beams 106
9.4 Practical examples 107
9.4.1 Raleigh Arena 107
9.4.2 Zhejiang Dragon Sports Centre 108
9.4.3 A cable-stayed bridge 111
9.4.4 A floor structure experiencing excessive vibration 111

10 Buckling 113

10.1 Definitions and concepts 113
10.2 Theoretical background 113
10.2.1 Buckling of a column with different boundary conditions 113
10.2.2 Lateral torsional buckling of beams 116
10.3 Model demonstrations 119
10.3.1 Buckling shapes of plastic columns 119
10.3.2 Buckling loads and boundary conditions 120
10.3.3 Lateral buckling of beams 122
10.3.4 Buckling of an empty aluminium can 123
10.4 Practical examples 124
10.4.1 Buckling of bracing members 124
10.4.2 Buckling of a box girder 125
10.4.3 Prevention of lateral buckling of beams 125

11 Prestress 127

11.1 Definitions and concepts 127
11.2 Theoretical background 127
11.3 Model demonstrations 133
11.3.1 Prestressed wooden blocks forming a beam and a column 133
11.3.2 A toy using prestressing 134
11.4 Practical examples 134
11.4.1 A centrally post-tensioned column 134
11.4.2 An eccentrically post-tensioned beam 135
11.4.3 Spider's web 135
11.4.4 A cable-net roof 137

12 Horizontal movements of structures induced by vertical loads 139

12.1 Concepts 139
12.2 Theoretical background 139
12.2.1 Static response 140
12.2.2 Dynamic response 149
12.3 Model demonstrations 151
12.3.1 A symmetric frame 151
12.3.2 An anti-symmetric frame 152

12.3.3 An asymmetric frame 152
12.4 Practical examples 153
    12.4.1 A grandstand 153
    12.4.2 A building floor 153
    12.4.3 Rail bridges 156

PART II
Dynamics   157

13   Energy exchange   159

13.1 Definitions and concepts 159
13.2 Theoretical background 159
13.3 Model demonstrations 164
    13.3.1 A moving wheel 164
    13.3.2 Collision balls 165
    13.3.3 Dropping a series of balls 167
13.4 Practical examples 168
    13.4.1 Rollercoasters 168
    13.4.2 A torch without a battery 169

14   Pendulums   170

14.1 Definitions and concepts 170
14.2 Theoretical background 170
    14.2.1 A simple pendulum 170
    14.2.2 A generalised suspended system 172
    14.2.3 Translational and rotational systems 176
14.3 Model demonstrations 176
    14.3.1 Natural frequency of suspended systems 176
    14.3.2 Effect of added masses 178
    14.3.3 Static behaviour of an outward inclined suspended system 180
14.4 Practical examples 182
    14.4.1 An inclined suspended wooden bridge in a playground 182
    14.4.2 Seismic isolation of a floor 182
    14.4.3 The Foucault pendulum 182

15   Free vibration   185

15.1 Definitions and concepts 185
15.2 Theoretical background 186
    15.2.1 A single-degree-of-freedom system 186
    15.2.2 A generalised single-degree-of-freedom system 191
    15.2.3 A multi-degrees-of-freedom (MDOF) system 195
    15.2.4 Relationship between the fundamental natural frequency and the maximum displacement of a beam 196

15.2.5 Relationship between the fundamental natural frequency and the tension force in a straight string 198

15.3 Model demonstrations 199
    15.3.1 Free vibration of a pendulum system 199
    15.3.2 Vibration decay and natural frequency 200
    15.3.3 An overcritically-damped system 201
    15.3.4 Mode shapes of a discrete model 202
    15.3.5 Mode shapes of a continuous model 202
    15.3.6 Tension force and natural frequency of a straight tension bar 203

15.4 Practical examples 204
    15.4.1 A musical box 204
    15.4.2 Measurement of the fundamental natural frequency of a building through free vibration generated using vibrators 206
    15.4.3 Measurement of the natural frequencies of a stack through vibration generated by the environment 207
    15.4.4 The tension forces in the cables in the London Eye 208

16 Resonance   210

16.1 Definitions and concepts 210
16.2 Theoretical background 210
    16.2.1 A SDOF system subjected to a harmonic load 211
    16.2.2 A SDOF system subject to a harmonic support movement 217
    16.2.3 Resonance frequency 219
16.3 Model demonstrations 221
    16.3.1 Dynamic response of a SDOF system subject to harmonic support movements 221
    16.3.2 Effect of resonance 222
16.4 Practical examples 222
    16.4.1 The London Millennium Footbridge 223
    16.4.2 Avoidance of resonance: design of structures used for pop concerts 225
    16.4.3 Measurement of the resonance frequency of a building 227
    16.4.4 An entertaining resonance phenomenon 228

17 Damping in structures   231

17.1 Concepts 231
17.2 Theoretical background 231
    17.2.1 Evaluation of viscous-damping ratio from free vibration tests 231
    17.2.2 Evaluation of viscous-damping ratio from forced vibration tests 233

17.3 Model demonstrations 234
    17.3.1 Observing the effect of damping in free vibrations 234
    17.3.2 Hearing the effect of damping in free vibrations 234
17.4 Practical examples 235
    17.4.1 Damping ratio obtained from free vibration tests 235
    17.4.2 Damping ratio obtained from forced vibration tests 237
    17.4.3 Reducing footbridge vibrations induced by walking 237
    17.4.4 Reducing floor vibration induced by walking 238

18 Vibration reduction 241

18.1 Definitions and concepts 241
18.2 Theoretical background 241
    18.2.1 Change of dynamic properties of systems 242
    18.2.2 Tuned mass dampers 244
18.3 Model demonstrations 246
    18.3.1 A tuned mass damper (TMD) 246
    18.3.2 A tuned liquid damper (TLD) 247
    18.3.3 Vibration isolation 248
18.4 Practical examples 248
    18.4.1 Tyres used for vibration isolation 248
    18.4.2 The London Eye 249
    18.4.3 The London Millennium Footbridge 249

19 Human body models in structural vibration 252

19.1 Concepts 252
19.2 Theoretical background 252
    19.2.1 Introduction 252
    19.2.2 Identification of human body models in structural vibration 254
19.3 Demonstration tests 257
    19.3.1 The body model of a standing person in the vertical direction 257
    19.3.2 The body model of a standing person in the lateral direction 259
19.4 Practical examples 261
    19.4.1 The effect of stationary spectators on a grandstand 261
    19.4.2 Calculation of the natural frequencies of a grandstand 263
    19.4.3 Dynamic response of a structure used at pop concerts 263
    19.4.4 Indirect measurement of the fundamental natural frequency of a standing person 263
    19.4.5 Indirect measurement of the fundamental natural frequency of a chicken 264

Index 266

Preface

According to the *Longman Dictionary of English Language and Culture*, a *concept* is defined as a *principle* or *idea*. A principle is defined as *a truth or belief that is accepted as a base for reasoning or action*. Structural concepts are amongst the main foundations for study, analysis and design in civil and structural engineering.

Structural concepts are key elements for students to understand, for lecturers to teach and for engineers to use in civil and structural engineering practice. The teaching of structural concepts at university needs to be enhanced to meet changes and challenges in our current learning environment and in the world of work.

In the past a major part of this understanding has been developed through working with hand calculations and through experience with construction. However, now many hand calculations are replaced by the use of computers and new methods of gaining an understanding of structural concepts are desirable. Indeed the understanding of structural concepts, fundamental to the sound and innovative design of structures (buildings, bridges, etc.) is even more important because of the wide use of computers and the often unquestioning reliance placed on the results of computer analyses which, although mathematically correct, may be flawed if they are based on incorrect assumptions and modelling. This is one reason for criticisms from the construction industry that graduates tend to place over reliance on the use of computers. Graduates, in general, are good at using computers but many are unable to judge whether the results obtained from computers are correct. This suggests that students may not have become adequately familiar with basic structural concepts during their university studies.

Structural concepts and principles are abstract, and they cannot be seen and felt directly. For instance, *force paths* transmit loads from their points of action to structural supports, and *resonance* describes the vibration characteristics of a structure responding to a dynamic load applied at a natural frequency of the structure. If such concepts and principles could be made more observable and touchable, students would be better able to understand and remember them.

Engineering examples are often not provided in textbooks to illustrate the applications of structural concepts. If lecturers could use related engineering examples and convert appropriate research work into teaching, the interest of students would be stimulated and their understanding of structural concepts would inevitably improve.

It has been observed in class situations that students show a greater interest in topics which are demonstrated physically than in topics that are explained by words and blackboard/OHP/PowerPoint presentations. They show an even greater interest in practical examples which illustrate the use of concepts in the solution of engineering problems rather than in coursework examples. Students are motivated by 'hands on' experience and by linking concepts and models to real engineering problems.

In such a background, we have been developing what we called *seeing and touching structural concepts* to supplement traditional class teaching and learning. To enable this, three parallel themes are followed:

- providing a series of simple demonstration models for illustrating structural concepts and principles in conventional class teaching which allow students to gain a better understanding of the concepts;
- providing associated engineering examples to demonstrate the application of the structural concepts and principles which help to bridge the gap between the students' knowledge and practice;
- converting appropriate research output, which particularly involves structural concepts, into teaching material to improve existing course contents.

Structural concepts that can be physically demonstrated are identified and simple demonstration models, suitable for class use, are provided to illustrate the concepts. Whenever possible, students have been encouraged to help to design and make these models.

Real interest can be generated and a better understanding can be achieved by seeing how concepts are used in the design of real structures. Therefore engineering examples which can illustrate the application of the concepts in practice have been sought and identified. Poor designs which may be illustrated by collapses have also been studied as such applications can often show the consequences of misunderstanding structural concepts.

Research and teaching are undertaken in parallel in universities but links between research output and undergraduate teaching may not always be developed. Research output, which particularly concerns or illustrates structural concepts, has been converted to forms suitable for linking with simple demonstration models and practical applications for use in class teaching. For example, concepts for designing stiffer structures, human whole-body models in structural vibration and the horizontal movements of frame structures induced by vertical loads are presented in the book.

We have developed a number of physical models for illustrating structural concepts and identified a number of engineering cases and everyday examples for illustrating the applications of these concepts. These models and examples are normally not included in textbooks but are useful to supplement learning and teaching. Students quickly grasp and remember a concept when it is physically demonstrated and its application is illustrated.

This book, which provides examples of links between structural concepts, simple demonstration models and engineering examples, will be an aid to lecturers and should benefit students and engineers/designers in structural engineering and architecture. It is hoped that it will stimulate the interest of readers to seek further examples of concepts and their applications.

This book is written like a 'recipe' book; most of the structural concepts in the book are independent and each chapter illustrates one or more related concepts. Each chapter contains four sections:

1. Definitions and concepts: definitions of the terms used in the chapter are provided. The concepts are presented concisely in one or two sentences and in a memorable manner. Key points are also given in some chapters.
2. Theoretical background: if the theory is readily available in textbooks, only a brief summary is presented together with appropriate references. More details are given when the theory is not readily available elsewhere. Selected examples are provided, which aim to show the use of the theory and link with the demonstration models illustrated in the next section.
3. Model demonstrations: demonstration models are provided with photographs. Normally, two related models are provided to show the differences between the behaviour of the two models thus illustrating the concept. Small-scale experiments are also included in some chapters.
4. Practical examples: appropriate engineering examples are given to show how the concept has been applied in practice. Some examples come from everyday life which should be familiar to most people.

Accompanying this book, we have created a website of the same title which can be found at www.structuralconcepts.org. The website contains most of the contents presented in the first, third and fourth sections of each chapter in this book. Colour photos can be downloaded from the website and video clips can be played.

The website and the book have been written initially for students, lecturers and graduate engineers in Civil and Structural Engineering. However, the contents may also be useful to similar groups in Architecture, Mechanical and Aerospace Engineering.

To students

This book provides useful and interesting information to enhance the understanding of structural concepts and to supplement class studies. The level of contents spans from first year to fourth year of a typical undergraduate course.

The contents of the book can be used in different ways:

1. You can look at any particular chapter after you have been introduced to a concept in the classroom to help you to gain a better understanding.
2. You can use the book to revise what you have learned in the past.
3. You may ask yourself if you can think of another model which is able to illustrate a concept listed (or indeed one that is not listed) or another example which shows the application of a concept.

Some of the models in the book were developed by our students as they knew which concepts were difficult to understand and which concepts could be physically demonstrated.

To lecturers

We hope the book provides useful material to supplement the teaching of structural concepts. The photos and/or contents in the associated website can be downloaded for teaching.

It is hoped that students learn effectively and actively and this, in part, requires the provision of appropriate activities and/or stimulators. This book and the website can be used for such a purpose. We have asked our third-year undergraduate students to read all the contents of the website relating to Statics and assigned them individual coursework entitled 'Enhancing the Understanding of Structural Concepts' for which they needed either to design a physical model to show one structural concept or to identify a practical example where one structural concept was creatively used. This coursework has been received enthusiastically and the returns have been excellent and collected to form a booklet. This was then distributed to all students in the class to form a student source of learning, enabling them to learn from each other rather than from lecturers and textbooks. Some examples provided by the students have been included in this book and added to the website.

To engineers

The contents of this book and the associated website are useful to engineers, in particular, recent graduate engineers. Unlike university students, you will have gained practical experience but may have forgotten some structural concepts you learned at university. You may find the book and website useful in three ways:

1. You can revise quickly many structural concepts.
2. You can examine the use of each of the concepts in practice through the examples provided. It is hoped that this may generate your own ideas for applying the concepts or indeed any other structural concepts in your work.
3. You can identify, using your own experience, how any of the structural concepts have been used in your work or in the work of your colleagues. Then consider how the application of the concepts helps in enhancing your understanding of structural behaviour and providing more efficient structures.

Engineers aim to achieve safe, economical and elegant designs. A good understanding of structural concepts will help to reach this goal.

Finally, we would very much like to hear from you if you have an idea to illustrate a structural concept using a physical model or have any practical examples in which one or more structural concepts plays an important role. We will acknowledge and add any suitable examples into the website and your contribution will be shared with others, including students, lecturers and practising engineers.

Tianjian Ji and Adrian Bell
School of Mechanical, Aerospace and Civil Engineering
University of Manchester, UK

Acknowledgements

Our work on this topic began in late 1999. We started identifying the concepts in textbooks for teaching undergraduates in Civil and Structural Engineering, which can be demonstrated using physical models and the real examples in engineering practice showing the application of the concepts. The work, however, could not be presented in the current form without the input of others.

Several previous and current undergraduate students at the University of Manchester, Miss W. Yip, Mr K. H. Lee, Mr C. E. Liang, Mr K. Chow, Mr K. Y. Chan and Mr T. Eccles, carried out investigative projects on Seeing and Touching Structural Concepts. They contributed their understanding through personal experience of study in class and made a number of physical models.

Mr M. Dean, Mr G. Lester, Mr G. Sigh and some other technical staff in the School gave assistance to these students with making models.

Professor Biaozhong Zhuang, an emeritus professor in Engineering Mechanics at Zhejiang University, China, contributed his personal experience of solving practical problems using basic theory of Mechanics and offered some interesting models used in daily life.

Several individuals and organisations kindly permitted the use of their photographs and they are acknowledged directly next to the photographs.

Mr T. Zheng, a PhD student at the University of Manchester, provided many of the line drawings in the book. Dr K. C. Leong, a previous student, and Miss L. Xue, a PhD student at the University of Manchester, helped to create the website.

We are also grateful to Dr B. R. Ellis, an independent consultant, for reading, checking and commenting on the manuscript.

The assistance of Taylor & Francis in the publication of this book is greatly appreciated. We would like to thank Tony Moore and Simon Bates for the encouragement and editorial assistance. We are also grateful for the help we received from Matt Deacon and his team at Wearset in preparing the final version of the manuscript.

Finally, we would like to acknowledge the financial support provided by The Education Trust of the Institution of Structural Engineers and the Faculty of Engineering and Physical Sciences at the University of Manchester for developing the website. The contents of the website form part of the book.

Part I
Statics

1 Equilibrium

1.1 Definitions and concepts

- A body which is not moving is in a state of equilibrium.
- A body is in a state of stable equilibrium when any small movement increases its potential energy so that when released it tends to resume its original position.
- A body is in a state of unstable equilibrium if, following any small movement, it tends to move away from its original position.
- The sum of reaction forces on a body is always equal to the sum of action forces if the body is in a state of equilibrium, regardless of the positions and magnitudes of the action forces.
- If a body is in an equilibrium state, all the moments acting on the body about any point should be zero.

1.2 Theoretical background

Rigid body: when considering equilibrium of an object subjected to a set of forces, the object is considered as a rigid body when its deformations do not affect the position of equilibrium and/or the forces (either directions or amplitudes) applied on the body.

Internal forces: internal forces act between particles within a body and occur in equal and opposite collinear pairs. The sum of the internal forces within a body will therefore be zero and need not be considered when considering the overall equilibrium of the body.

External forces: external forces act on the surfaces of bodies and may be considered to be distributed and applied over contact areas. When the contact areas are very small in comparison to the area of the surface, the distributed forces can be idealised as single concentrated forces acting at points on the surface of the body. A typical example of an external force which is distributed over a surface is wind load acting on the walls of a building. Concentrated external forces may be considered to be applied to a bridge deck by a vehicle whose weight is transmitted to the deck through its wheels which have small contact areas relative to the area of the deck.

Moment: a moment can be either an internal or an external force that tends to cause rotation of a body. A pair of parallel forces with the same magnitude but opposite directions form a moment or couple. The magnitude of the moment is measured by the product of the magnitude of the forces and the perpendicular distance between the two parallel forces.

4 Statics

Body force: a body force, or the self-weight of a body, is the effect of Earth's gravity on the body. This force is not generated through direct contact like most other external forces. The self-weight of a body can be significant in the design of structures, for example, the self-weight of a concrete floor can be larger than other applied imposed loads such as those due to people or furnishings. Body forces, other than self-weight, can also be generated remotely by magnetic or electric fields (section 1.3.4)

Reactions: the forces developed in the supports or points of contact between bodies are called reactions. Supports are provided to prevent the movement of a body. Different types of support prevent different types of movement. The main types of support which occur in practice in plane structures are:

- **Pin supports:** a pin support restrains translational movements in two perpendicular directions but allows rotation (Figure 1.1a).
- **Roller supports:** a roller support restrains translational movement in one direction but allows translational movements in a perpendicular direction as well as allowing rotation (Figure 1.1b).
- **Fixed supports:** a fixed support restrains translational movements in two perpendicular directions and prevents rotation (Figure 1.1c).

Conditions of equilibrium of a rigid body: a body is in a state of equilibrium when it does not move. In this state, the effects of all the forces F and moments M applied to the body cancel each other. Considering all the forces and moments applied in the x–y plane, the conditions of equilibrium of the body can be specified using three scalar equations of equilibrium:

$$\sum F_X = 0 \quad \sum F_Y = 0 \quad \sum M_A = 0 \tag{1.1}$$

Equation 1.1 indicates that:

- The sums of all the forces in the x and y directions are zero and the sums of the moments induced by these forces about an arbitrary point are zero.
- The three equations of equilibrium can be used to determine three unknowns in an equilibrium problem, such as the reaction forces of a simple body in the x–y plane.

Free-body diagram: a free-body diagram is a diagrammatic representation of a body or a part of a body showing all forces applied on it by mechanical contact with other bodies or parts of the body itself. Drawing a free-body diagram which shows

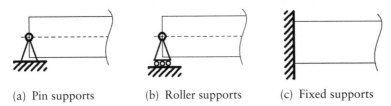

(a) Pin supports   (b) Roller supports   (c) Fixed supports

Figure 1.1 Examples of support conditions.

clearly and completely all the forces acting on the body is an important and necessary skill.

Example 1.1

Figure 1.2a shows a beam which is supported at the left-hand end on a pin support and at the right-hand end on a roller support. Such a beam is called a *simply supported beam*. The beam carries a concentrated vertical force W acting at a distance x from the left-hand end of the beam. Determine the reaction forces, A_x, A_y and B_y, at the two supports.

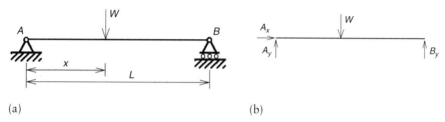

Figure 1.2 A simply supported beam with (a) a concentrated load and (b) its free-body diagram.

Solution

A free-body diagram of the beam is shown in Figure 1.2b where the external force and all the support forces are indicated. Using the three formulae in equation 1.1 gives:

$$\sum F_X = 0 \quad A_X = 0$$

$$\sum F_Y = 0 \quad A_Y + B_Y - W = 0$$

$$\sum M_A = 0 \quad B_Y L - W_X = 0$$

Solving the last two equations leads to: $A_Y = W - Wx/L$ and $B_Y = Wx/L$. The positive signs of the reaction forces indicate that A_Y and B_Y act in the assumed directions as shown in Figure 1.2b. The second formula shows that *the sum of the reaction forces in the y direction is always equal to the external force W, no matter where it is placed*. A model demonstration of this example is given in section 1.3.1. Further details can be seen in [1.1].

1.3 Model demonstrations

1.3.1 Action and reaction forces

This model demonstrates *what has been shown in example 1.1, namely*:

- *The sum of reaction forces is always equal to the external force, W, regardless of the location of the external force.*

6 Statics

- *When the external force, W, is placed above one of the two supports, the reaction force at that support is equal to the external force but acts in the opposite direction and there is no reaction force at the other support.*

Figure 1.3a shows a wooden beam supported by two scales, one at each end. The scales are adjusted to zero when the beam is in place. Locate a weight of 454 grams at three different positions on the beam and note the readings on the two scales. It can be seen that:

- For the three positions of the weight, the sum of the readings (reactions) from the two scales is always equal to the weight (the external force), see Figure 1.3b.
- The readings on the scales have a linear relation to the distance between the locations of the supports and the weight. The closer the weight is to one of the scales, the larger the reading on the scale. An extreme case occurs when the weight is placed directly over the left-hand scale. The reading on the left-hand scale is the same as the weight and the reading on the other scale is zero (Figure 1.3c).

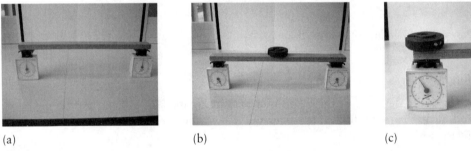

(a) (b) (c)

Figure 1.3 Action and reaction.

1.3.2 Stable and unstable equilibrium

This model demonstration shows *the difference between stable and unstable equilibrium.*

The equilibrium of a ruler supported on two round pens (or roller supports) can be achieved easily as shown in Figure 1.4a. However, supporting the ruler horizontally on a single round pen alone is very difficult, because the external force (self-

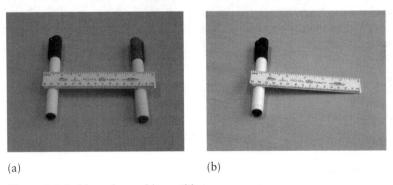

(a) (b)

Figure 1.4 Stable and unstable equilibrium.

weight) from the ruler and the reaction force from the round pen are difficult to align exactly. If the ruler achieves equilibrium on the single round pen, this type of equilibrium is unstable and is not maintained if a slight disturbance is applied to the ruler or to the round pen. The ruler will rotate around the point of contact with the pen until it finds another support (Figure 1.4b).

1.3.3 A plate–bottle system

This model demonstration shows *how an equilibrium state can be achieved.*

Figure 1.4 shows that the ruler is in stable equilibrium when it has two round pen supports and becomes unstable equilibrium when it has only one round pen support. This, however, does not mean that a body placed on a single support cannot achieve a state of stable equilibrium.

Figure 1.5a shows a bottle of wine and a piece of wood with a hole. The bottle can be supported by the wood when the neck of the bottle is inserted into the hole to the maximum extent, and the two form a single wood–bottle system in equilibrium as shown in Figure 1.5b.

(a)   (b)

Figure 1.5 Equilibrium of a bottle and wood system.

The wood–bottle system, supported on the narrow wooden edge, is and feels very stable, because:

- the two external forces from the weights of the bottle and the wood are equal to the reaction force generated from the table;
- the sum of the moments of the two action forces about the point where the support force acts are equal to zero.

Another way of saying this is that the centre of mass of the wood and bottle system lies over the base of the wood.

1.3.4 A magnetic 'float' model

The model demonstrates *the effect of magnetic force although the force cannot be seen.*

Figure 1.6 shows a model consisting of an axisymmetric body and a base unit.

8 Statics

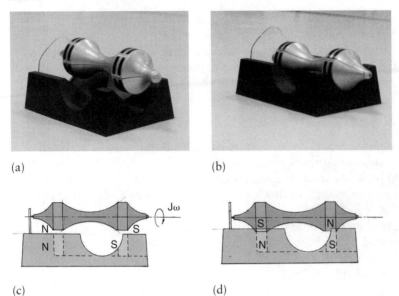

Figure 1.6 A magnetic float model (courtesy of Professor B. Zhuang, Zhejiang University, China).

There are two magnetic rings in the axisymmetric body and two magnets in the base unit.

It appears surprising that the axisymmetric body can be in an equilibrium position when no vertical supports are provided. Where is the force that supports the weight of the axisymmetric body? When opposite poles of the magnets in the axisymmetric body and the base unit are the same, the body can be positioned above the base with no visible support (Figures 1.6a and 1.6c). When the opposite poles are different, the body rests on the base (Figures 1.6b and 1.6d). In both cases the external force is the weight of the body. In the first case the reaction forces are the magnetic forces that push the body away and up while in the second case the reaction forces are provided through the points of contact between the unit and the base.

When the free end of the body (Figure 1.6a) is twisted using the thumb and index finger, the body is able to rotate many times before stopping. This is because there is little friction between the rotating body and its lateral glass support.

1.4 Practical examples

1.4.1 A barrier

Counter-balanced barriers can be found in many places, such as the one shown in Figure 1.7. The weight fixed to the shorter arm counter-balances much of the weight of the barrier arm, allowing the barrier to be opened easily with a relatively small downward force near the counter-balance weight.

The barrier can be opened when the moment induced by the counter-balance weight and the applied downward force is larger than the moment induced by the weight of the barrier about the support point.

Equilibrium 9

Figure 1.7 A barrier.

1.4.2 A footbridge

The form of footbridge shown in Figure 1.8 is a development of the simple counterbalanced barrier shown in Figure 1.7.

Figure 1.8 A footbridge can be lifted by a single person.

10 Statics

The moment induced by the weight of the footbridge deck about the supporting points on the wooden frame is slightly larger than that induced by the balance weight, which is placed on and behind a wooden board. The addition of a small force, applied by pulling on a cable, causes the bridge to open.

1.4.3 An equilibrium kitchen scale

Figure 1.9 shows a kitchen scale that weighs an apple and a banana using the principle of equilibrium.

Figure 1.9 An equilibrium kitchen scale.

There is a pivot at the centre of the scale and a pair of arms can rotate about the pivot. The left and right arms are designed to be symmetrical about the central pivot. There are two other pivots at the ends of the left and right arms where two trays are placed. The two arms would be at the same level when an equilibrium state is achieved. The equilibrium equation for moment (the third equation in equation 1.1) can be used to give:

Weight of the fruits $\times d_L$ = standard weights placed $\times d_R$

where d_L and d_R are the distances between the central pivot and the left- and right-hand pivots of application of the weights. Due to the symmetry, $d_L = d_R$, the above equation thus states that the weight of the fruits is equal to the standard weights placed on the tray. The operation of the scale is thus based on the principle of equilibrium.

1.4.4 Stage performance

Equilibrium in the posture of the human body can be important in sport and dance.

The two people shown in Figure 1.10 have positioned themselves horizontally

Equilibrium 11

Figure 1.10 Equilibrium on stage (photograph of Crazeehorse, copyright of Fremantle Media, reproduced by kind permission).

such that the centre of mass of their bodies lies over the feet of one of the performers. The balance shows the strength of the performers and the beauty of the posture.

1.4.5 Magnetic float train

Figure 1.11 shows the first commercial friction-free magnetic 'float' train, which operates between Shanghai City and Pudong Airport. The floating train can reach speeds in excess of 270 mph. The train, with magnets under the cars, 'floats' above a sophisticated electromagnetic track in the same way that the axisymmetric object 'floats' above the base unit in the demonstration in section 1.3.4. The weight of the train (external force) is balanced by magnetic forces (reaction forces) generated between the magnets in the cars and the track. Since no friction is generated between the external and reaction forces, much smaller tractive forces are required to propel the train than would normally be the case.

Float trains have also run in the UK on a prototype route between Birmingham Airport and a nearby train station, but they were eventually abandoned in 1995 because of reliability problems.

1.4.6 A dust tray

Equilibrium may also be found in several common items used in daily life. Figure 1.12a shows a dust tray which is often used in fast-food restaurants to collect rubbish from floors. When moving the dust tray between locations, one lifts the vertical handle. The tray then rotates through an angle such that the rubbish will be kept at the end of the tray as shown in Figure 1.12b. The rotation occurs because

Figure 1.11 A magnetic float train.

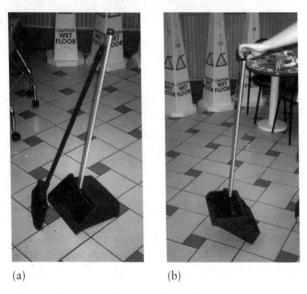

(a)　　　　　　　　　(b)

Figure 1.12 An application of inequilibrium.

there is a pivot on the tray and the pivot is purposely placed at the position which will cause the tray to turn. Because the mass centre of the dust tray in its horizontal position is not in line with the handle, the tray rotates until this is the case. If the larger portion of the tray lies behind the handle, the direction of rotation is such that rubbish remains trapped during movement.

Reference

1.1 Hibbeler, R. C. (2005) *Mechanics of Materials*, Sixth Edition, Singapore: Prentice-Hall Inc.

2 Centre of mass

2.1 Definitions and concepts

The centre of gravity of a body is the point about which the body is balanced or the point through which the weight of the body acts.

The location of the centre of gravity of a body coincides with the **centre of mass of the body** when the dimensions of the body are much smaller than those of the earth.

When the density of a body is uniform throughout, the centre of mass and the **centroid of the body** are at the same point.

- If the centre of mass of a body is not positioned above the area over which it is supported, the body will topple over.
- The lower the centre of mass of a body, the more stable the body.

2.2 Theoretical background

Centre of gravity: the centre of gravity of a body is a point through which the weight of the body acts or is a point about which the body can be balanced. The location of the centre of gravity of a body can be determined mathematically using the moment equilibrium condition for a parallel force system. Physically, a body is composed of an infinitive number of particles and if a typical particle has a weight of dW and is located at (x, y, z) as shown in Figure 2.1, the position of the centre of gravity, G, at $(\bar{x}, \bar{y}, \bar{z})$ can be determined using the following equations:

$$\bar{x} = \frac{\int x\, dW}{\int dW} \quad \bar{y} = \frac{\int y\, dW}{\int dW} \quad \bar{z} = \frac{\int z\, dW}{\int dW} \tag{2.1}$$

Centre of mass: to study problems relating to the motion of a body subject to dynamic loads, it is necessary to determine the centre of mass of the body. For a particle with weight dW, and mass dm, the relationship between mass and weight is given by $dW = g\, dm$, where g is the acceleration due to gravity. Substituting this relationship into equation 2.1 and cancelling g from both the numerator and denominator leads to:

$$\bar{x} = \frac{\int x\, dm}{\int dm} \quad \bar{y} = \frac{\int y\, dm}{\int dm} \quad \bar{z} = \frac{\int z\, dm}{\int dm} \tag{2.2}$$

14 Statics

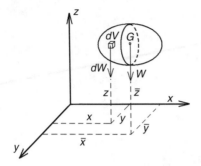

Figure 2.1 Centre of gravity.

Equation 2.2 contains no reference to gravitational effects (*g*) and therefore defines a unique point which is a function solely of the distribution of mass. This point is called the centre of mass of the body. For civil engineering structures whose dimensions are very small compared to the dimensions of the earth, *g* may be considered to be uniform and hence the centre of mass of a body coincides with the centre of gravity of the body. In practice, no distinction is made between the centre of mass and the centre of gravity of a body.

The mass, dm, of a particle in a body relates to its volume, dV, and can be expressed as $dm = \rho(x, y, z)dV$, where $\rho(x, y, z)$ is the density of the body. Substituting this relationship into equation 2.2 gives:

$$\bar{x} = \frac{\int x \rho(x, y, z)dV}{\int \rho(x, y, z)dV} \quad \bar{y} = \frac{\int y \rho(x, y, z)dV}{\int \rho(x, y, z)dV} \quad \bar{z} = \frac{\int z \rho(x, y, z)dV}{\int \rho(x, y, z)dV} \tag{2.3}$$

Centroid: the centroid C of a body is a point which defines the geometrical centre of the body. If a body is composed of one type of material, or the density of the material ρ is constant, it can be removed from equation 2.3 and the centroid of the body can be defined by:

$$\bar{x} = \frac{\int x\,dV}{\int dV} \quad \bar{y} = \frac{\int y\,dV}{\int dV} \quad \bar{z} = \frac{\int z\,dV}{\int dV} \tag{2.4}$$

Equation 2.4 contains no reference to the mass of the body and therefore defines a point, the centroid, which is a function solely of geometry.

If the density of the material of a body is uniform, the locations of the centroid and the centre of mass of the body coincide. However, if the density varies throughout a body, the centroid and the centre of mass will not be the same in general (section 2.3.2).

Equation 2.4 is used to determine the centroid of the volume of a body. The volume of a particle of the body, dV, can be expressed as $dV = A\,dt$ where A is the area of the particle perpendicular to the axis through the thickness, dt, of the particle.

When the centroid of a surface area of an object, such as a flat plate, needs to be determined, equation 2.4 reduces to:

$$\bar{x} = \frac{\int x\,dA}{\int dA} \quad \bar{y} = \frac{\int y\,dA}{\int dA} \quad \bar{z} = \frac{\int z\,dA}{\int dA} \tag{2.5}$$

The terms such as $x\,dA$ in the numerators in the integrals represent the moments of the area of the particle, dA, about the centroid. The integrals represent the summations of these moments of areas for the whole body and are usually called the first moments of area of the body about the axes in question. The integral in the denominators in equation 2.5 is the area of the body.

When the centroid of a line, effectively a body with one dimension, needs to be determined, equation 2.5 can be simplified to:

$$\bar{x} = \frac{\int x\,dL}{\int dL} \quad \bar{y} = \frac{\int y\,dL}{\int dL} \quad \bar{z} = \frac{\int z\,dL}{\int dL} \tag{2.6}$$

where dL is the length of a short element or part of the line.

Equations 2.4, 2.5 and 2.6 are based on a coordinate system that may be chosen to suit a particular situation and in general such a coordinate system should be chosen to simplify as much as possible the integrals. When the geometric shape of an object is simple, the integrals in the equations can be replaced by simple algebraic calculations.

Example 2.1

Determine the centroid of the uniform L-shaped object shown in Figure 2.2a.

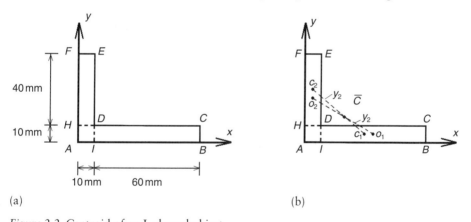

(a)   (b)

Figure 2.2 Centroid of an L-shaped object.

Solution 1

As the L-shaped object consists of two rectangles, $ABCH$ and $HDEF$, these areas and the moments of the areas about the axes can be easily calculated without the need for integration. The areas and their centroids for the coordinate system chosen are:

$ABCH$: $A_1 = 70 \times 10 = 700 \text{ cm}^2$, $\quad x_{C1} = 35 \text{ cm}, \quad y_{C1} = 5 \text{ cm}$

$HDEF$: $A_2 = 40 \times 10 = 400 \text{ cm}^2$, $\quad x_{C2} = 5 \text{ cm}, \quad y_{C2} = 30 \text{ cm}$

16 Statics

Using the first two formulae in equation 2.5 gives:

$$\bar{x} = \frac{A_1 x_{C1} + A_2 x_{C2}}{A_1 + A_2} = \frac{700 \times 35 + 400 \times 5}{700 + 400} = \frac{26500}{1100} = 24.09\,\text{cm}$$

$$\bar{y} = \frac{A_1 y_{C1} + A_2 y_{C2}}{A_1 + A_2} = \frac{700 \times 5 + 400 \times 30}{700 + 400} = \frac{15500}{1100} = 14.09\,\text{cm}$$

Solution 2

The centroid of the L-shaped object can also be determined using a geometrical method. The procedure can be described as follows:

- Divide the L shape into two rectangles, *ABCH* and *HDEF*, as shown in Figure 2.2b. Find the centroids of the two rectangles, C_1 and C_2, and then draw a line linking C_1 and C_2. The centroid of the object must lie on the line C_1–C_2. The equation of this line is $y - 5 = -5(x - 35)/6$.
- Divide the shape into two other rectangles, *IBCD* and *AIEF* as shown in Figure 2.2b. Find the centres of the two rectangles, O_1 and O_2, then draw a line joining their centres. The centroid of the object must also lie on the line O_1–O_2 which is $y - 5 = -4(x - 40)/7$.
- As the centroid of the object lies on both lines C_1–C_2 and O_1–O_2, it must be at their intersection, $\bar{C}$, as shown in Figure 2.2b. Solving the two simultaneous equations for x and y leads to $\bar{x} = 24.09$ cm and $\bar{y} = 14.09$ cm.

It can be seen from Figure 2.2b that the centroid, $\bar{C}$, is located outside the boundaries of the object. As the density of the L-shaped object is uniform, the centre of gravity, the centre of mass and the centroid coincide, so all lie outside the boundaries of the object. From the definitions of centre of mass and centre of gravity given in section 2.1, the L-shaped object can be supported by a force applied at the centre of mass, perpendicular to the plane of the object. In practice, this may be difficult as the centre of mass lies outside the object. Section 2.3.3 provides a model demonstration that shows how an L-shaped body can be lifted at its centre of mass.

Example 2.2

Determine the centre of mass of a solid, uniform pyramid that has a square base of

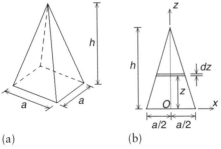

Figure 2.3 Centroid of a pyramid.

$a \times a$ and a height of h. The base of the pyramid lies in the x–y plane and the apex of the pyramid is aligned with the z axis as shown in Figure 2.3a.

Solution

Due to symmetry the centroid must lie on the z axis, i.e.

$$\bar{x} = \bar{y} = 0$$

The location of the centroid of the pyramid on the z axis can be obtained by using the third formula in equation 2.4 where the integral is taken over the volume of the pyramid. Consider a slice of thickness dz of the pyramid which is parallel to the base with a distance z from the base as shown in Figure 2.3b. The length of the side of the square slice is $a(1-z/h)$, and hence the volume of the slice is $dV = A(z)dz = a^2(1-z/h)^2 dz$. Substituting the above quantities into the third formula in equation 2.4 and integrating with respect to z through the height of the pyramid gives:

$$z_c = \frac{\int_0^h za^2(1-z/h)^2 dz}{\int_0^h a^2(1-z/h)^2 dz} = \frac{a^2 h^2/12}{a^2 h/3} = \frac{h}{4}$$

Thus the centre of mass of a uniform square-based pyramid is located on its vertical axis of symmetry at a quarter of its height measured from the base.

Example 2.3

The location of the centre of mass of a uniform block is at a height of h from its base that has a side length of b. Place the block on an inclined surface as shown in Figure 2.4 and gradually increase the slope of the surface until the block topples over. Determine the maximum slope for which the block does not topple over (assuming there is no sliding between the base of the block and the supporting surface).

Solution

It would normally be assumed that the block would not topple over if the action line of its weight remains within the area of the base of the block. Thus the critical position for the block is when the centre of gravity of the block and the corner of

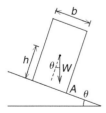

Figure 2.4 Centre of mass and stability.

18 *Statics*

the base lie in the same vertical line. Therefore, the critical angle of tilt for the block is:

$$\theta_C = \tan^{-1}\frac{b/2}{h} = \tan^{-1}\frac{b}{2h} \tag{2.7}$$

If the assumption is correct, equation 2.7 indicates that if $\theta < \theta_C$ the block will not topple over. The assumption can be checked by conducting the simple model tests in section 2.3.5. Further details can be seen in [2.1].

2.3 Model demonstrations

2.3.1 *Centre of mass of a piece of cardboard of arbitrary shape*

This demonstration shows how *the centre of mass of a body with arbitrary shape can be determined experimentally.*

Take a piece of cardboard of any size and shape, such as the one shown in Figure 2.5, and drill three small holes at arbitrary locations along its edge. Now suspend the cardboard using a drawing pin through one of the holes and hang from the drawing pin a length of string supporting a weight. As the diameter of the hole is larger than the diameter of the pin, the cardboard can rotate around the pin and will be in equilibrium under the action of the self-weight of the cardboard. This means that the resultant of the gravitational force on the cardboard must pass through the pin. In other words, the action force (gravitational force) and the reaction force from the pin must be in a vertical line which is shown by the string. Now mark a point on the cardboard under the string and draw a straight line between this point and the hole from which the cardboard is supported. Repeat the procedure hanging the cardboard from each of the other two holes in turn and in each case draw a line between a point on the string and the support hole. It can be seen that the three lines (or their extensions) are concurrent at a single point, which is the centre of mass of

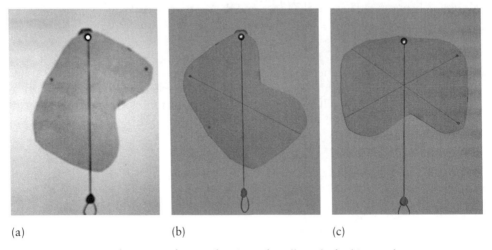

(a) (b) (c)

Figure 2.5 Locating the centre of mass of a piece of cardboard of arbitrary shape.

the piece of cardboard and of course is also the centre of gravity and the centroid of the cardboard.

2.3.2 Centre of mass and centroid of a body

This model is designed to show that *the location of the centre of mass of a body can be different from that of the centroid of the body.*

A 300 mm tall pendulum model is made of wood. The model consists of a supporting base, a mast fixed at the base and a pendulum pinned at the other end of the mast as shown in Figure 2.6. A metal piece is inserted into the right arm of the pendulum. The centre of mass of the pendulum must lie on a vertical line passing through the pinned point. The centroid of the pendulum lies on the intersection of the vertical and horizontal members. Due to the different densities of wood and metal, the centroid of the pendulum does not coincide with the centre of mass of the body.

Figure 2.6 Centre of mass and centroid.

2.3.3 Centre of mass of a body in a horizontal plane

This demonstration shows *how to locate the centre of mass of a horizontal L-shaped body.*

Figure 2.2b shows an L-shaped area whose centre of mass, C, is outside the body. The following simple experiment, using a fork, a spoon, a toothpick and a wine glass, can be carried out to locate the centre of mass of an L-shaped body made up from a spoon and fork.

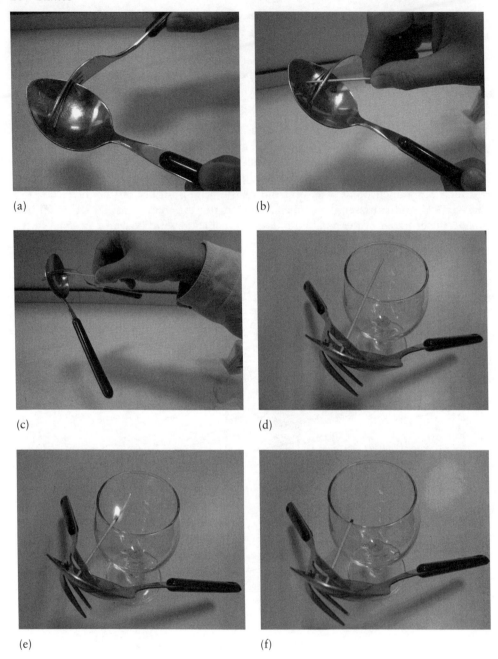

(a) (b) (c) (d) (e) (f)

Figure 2.7 Centre of mass of an L-shaped body in a horizontal plane.

1   Take a spoon and a fork and insert the spoon into the prongs of the fork, to form an L-shape as shown in Figure 2.7a.
2   Then take a toothpick and wedge the toothpick between two prongs of the fork and rest the head of the toothpick on the spoon as shown in Figure 2.7b. Make sure the end of the toothpick is firmly in contact with the spoon.

Centre of mass   21

3 The spoon–fork can then be lifted using the toothpick. The toothpick is subjected to an upward force from the spoon, a downward force from the prong of the fork and forces at the point at which it is lifted as shown in Figure 2.7c.
4 Place the toothpick on to the edge of a wine glass adjusting its position until the spoon–fork–toothpick is balanced at the edge of the glass as shown in Figure 2.7d. According to the definition of the centre of mass, the contact point between the edge of the glass and the toothpick is the centre of mass of the spoon–fork–toothpick system. The mass of the toothpick is negligible in comparison to that of the spoon–fork, so the contact point can be considered to be the centre of mass of the L-shaped spoon–fork system.
5 To reinforce the demonstration, set alight the free end of the toothpick (Figure 2.7e). The flame goes out when it reaches the edge of the glass as the glass absorbs heat. As shown in Figure 2.7f, the spoon–fork just balances on the edge of the glass.

If the spoon and fork are made using the same material, the centre of mass and the centroid of the spoon–fork system will coincide.

2.3.4 Centre of mass of a body in a vertical plane

This demonstration shows that *the lower the centre of mass of a body, the more stable the body*.

A cork, two forks, a toothpick, a coin and a wine bottle are used in the demonstration shown in Figure 2.8a. The procedure for the demonstration is as follows:

1 Push a toothpick into one end of the cork to make a toothpick–cork system as shown in Figure 2.8b.
2 Try to make the system stand up with the toothpick in contact with the surface of a table. In practice this is not possible as the centre of mass of the system is high and easily falls outside the area of the support point. The weight of the

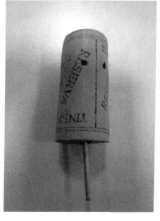

(a)   (b)   (c)

Figure 2.8 Centre of mass of a body in a vertical plane.

22 Statics

toothpick–cork and the reaction from the surface of the table that supports the toothpick–cork are not in the same vertical line and this causes overturning of the system.

3   Push two forks into opposite sides of the cork and place the coin on the top of the bottle.
4   Place the toothpick–cork–fork system on the top of the coin on which the toothpick will stand in equilibrium as shown in Figure 2.8c.

The addition of the two forks significantly lowers the centre of mass of the toothpick–cork system in the vertical plane below the contact point of the toothpick on the coin. When the toothpick–cork–fork system is placed on the coin at the top of the bottle, the system rotates until the action and reaction forces at the contact point of the toothpick are in the same line, producing an equilibrium configuration.

2.3.5 Centre of mass and stability

This demonstration shows how *the stability of a body relates to the location of its centre of mass and the size of its base.*

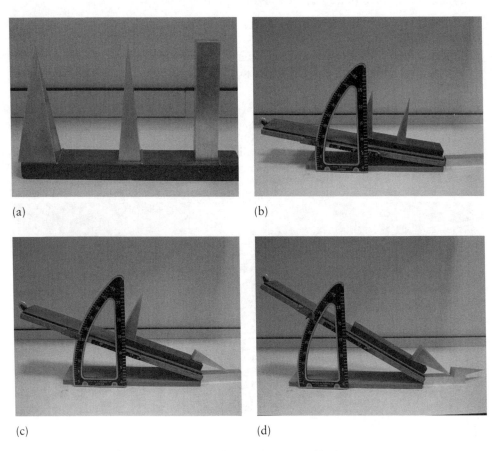

Figure 2.9 Centre of mass and stability of three aluminium blocks.

Figure 2.9a shows three aluminium blocks with the same height of 150 mm. The square sectioned block and the smaller pyramid have the same base area of 29 mm × 29 mm. The larger pyramid has a base area of 50 mm × 50 mm but has the same volume as that of the square sectioned block. The three blocks are placed on a board with metal stoppers provided to prevent sliding between the base of the blocks and the board when the board is inclined. As the board is inclined, its angle of inclination can be measured by the simple equipment shown in Figure 2.9b. Basic data for the three blocks and the theoretical critical angles calculated using equation 2.7 are given in Table 2.1. Theory predicts that the largest critical angle occurs with the large pyramid and the smallest critical angle occurs with the square sectioned block.

The demonstration is as follows:

1. The blocks are placed on the board as shown in Figure 2.9 in the order of increasing predicted critical angle.
2. The left-hand end of the board is gradually lifted and the square sectioned block is the first to become unstable and topple over (Figure 2.9b). The angle at which the block topples over is noted.
3. The board is inclined further and the pyramid with the smaller base is the next to topple (Figure 2.9c). Although the height of the centre of mass of the two pyramids is the same, the smaller pyramid has a smaller base and the line of action of its weight lies outside the base at a lower inclination than is the case for the larger pyramid. The angle at which the smaller pyramid topples over is noted.
4. As the board is inclined further the larger pyramid will eventually topple over but its improved stability over the other two blocks is apparent (Figure 2.9d). Once again the angle at which the block topples is noted.

The angles at which the three blocks toppled are shown in Table 2.1. The results of the demonstration as given in Table 2.1 show that:

- The order in which the blocks topple is as predicted by equation 2.7 in terms of the measured inclinations and confirms that the larger the base or the lower the centre of mass of a block, the larger the critical angle that is needed to cause the block to topple.
- All the measured critical angles are slightly smaller than those predicted by equation 2.7.

Repeating the experiment several times confirms the measurements and it can be observed that the bases of the blocks just leave the supporting surface immediately

Table 2.1 Comparison of the calculated and measured critical angles

Model	Cuboid	Small pyramid	Large pyramid
Height of the model (mm)	150	150	150
Height of the centre of mass (mm)	75	37.5	37.5
Width of the base (mm)	29	29	50
Volume (mm^3)	1.26×10^5	0.420×10^5	1.25×10^5
Theoretical max. inclination (deg.)	10.9	21.9	33.7
Measured max. inclination (deg.)	10	19	31

24 *Statics*

before they topple, which makes the centres of the masses move outwards. In the theory the bases of the blocks remain in contact with the support surface before they topple.

This demonstration shows that *the larger the base and/or the lower the centre of gravity, the larger will be the critical angle needed to cause the block to topple and that this angle is slightly less than that predicted by theory.*

2.3.6 *Centre of mass and motion*

This demonstration shows that *a body can appear to move up a slope unaided.*

Take two support rails which incline in both the vertical and horizontal planes as shown in Figure 2.10. When a doubly conical solid body is placed on the lower ends of the rails, it can be observed that the body rotates and travels up to the higher end of the rails (Figure 2.10b).

It appears that the body moves against gravity though in fact it moves with gravity. When the locations of the centre of mass of the body, C, are measured at the lowest and highest ends of the rails, it is found that the centre of mass of the body at the lower end is actually higher than that when the body is at the highest end of the rails. It is thus gravity that makes the body rotate and appear to move up the slope.

The reason for this lies in the design parameters of the rail supports and the conical body. The control condition is that the slope of the conical solid body should be larger than the ratio of the increased height to a half of the increased width of the rails between the two ends.

2.4 Practical examples

2.4.1 *Cranes on construction sites*

Tower cranes are common sights on construction sites. Such cranes normally have large weights placed on and around their bases, these weights ensuring that the centre of mass of the crane and its applied loading lie over its base area and that the centre

(a)

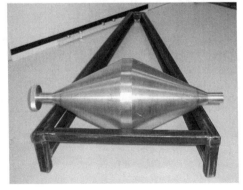
(b)

Figure 2.10 Centre of mass and motion.

Centre of mass 25

(a) (b)

Figure 2.11 Cranes on construction sites.

of mass of the crane is low, increasing its stability. Figure 2.11a shows a typical tower crane on a construction site. Concrete blocks are purposely placed on the base of the crane to lower its centre of mass to keep the crane stable as shown in Figure 2.11b.

2.4.2 The Eiffel Tower

Building structures are often purposely designed to be larger at their bases and smaller at their upper parts so that the distribution of the mass of the structure reduces with height. This lowers the centre of mass of the structure and the greater base dimensions reduce the tendency of the structure to overturn when subjected to lateral loads, such as wind. A typical example of such a structure is the Eiffel Tower in Paris as shown in Figure 2.12. The form of the tower is reassuring and appears to be stable and safe.

Figure 2.12 The Eiffel Tower.

26 *Statics*

2.4.3 A display unit

Figure 2.13 shows an inclined display unit. The centre of mass of the unit is outside the base of the unit as shown in Figure 2.13a. To make the unit stable and prevent it from overturning, a support base is fixed to the lower part of the display unit (Figure 2.13b). The added base effectively increases the size of the base of the unit to ensure that the centre of mass of the display unit will be above the area of the new base of the unit. Thus the inclined display unit becomes stable. Other safety measures are also applied to ensure that the display unit is stable.

(a) (b)

Figure 2.13 Inclined display unit.

2.4.4 The Kio Towers

Figure 2.14 shows one of the two 26-storey, 114 m high buildings of the Kio Towers, in Madrid, which are also known as Puerta de Europa (Gateway to Europe). The Kio Towers actually lean towards each other, each inclined at 15 degrees from the vertical.

The inclinations move the centres of mass of the buildings sideways, tending to cause toppling effects on the buildings. One of the measures used to reduce these toppling effects was to add massive concrete counterweights to the basements of the buildings. This measure not only lowered the centres of the mass of each building but also moved the centre of the mass towards a position above the centre of the base of the building.

Figure 2.14 One of the two Kio Towers.

Reference

2.1 Hibbeler, R. C. (2005) *Mechanics of Materials*, Sixth Edition, Singapore: Prentice-Hall Inc.

3 Effect of different cross-sections

3.1 Definitions and concepts

Second moment of area (sometimes incorrectly called moment of inertia) is the geometrical property of a plane cross-section which is based on its area and on the distribution of the area.

- The further the material of the section is away from the neutral axis of the section, the larger the second moment of area of the section. Therefore the shape of a cross-section will significantly affect the value of the second moment of area.
- The stiffness of a beam is proportional to the second moment of area of the cross-section of the beam.

3.2 Theoretical background

Second moment of area: the second moment of area of a section can be determined about its x and y axes from the following integrals:

$$I_x = \int y^2 dA \quad I_y = \int x^2 dA \tag{3.1}$$

where x and y are the coordinates of the elemental area dA and are measured from the neutral axes of the section. The use of these equations can be illustrated by the simple example of the rectangular cross-section shown in Figure 3.1, where the x and y axes have their origin at the centroid C, and b and h are the width and height of the section. Consider a strip of width b and thickness dy which is parallel to the x axis and a distance y from the x axis. The elemental area of the strip is $dA = bdy$. Using the first expression in equation 3.1 gives the second moment of area of the section about the x axis as:

$$I_x = \int_{-h/2}^{h/2} y^2 b\, dy = \frac{bh^3}{12} \tag{3.2}$$

It can be seen from equation 3.2 that I_x is proportional to the width b and to the height h to the power of three, i.e. increasing the height h is a much more effective way of increasing the second moment of area, I_x, than increasing the width b.

Similarly, consider a strip of width h and thickness dx which is parallel to the y axis and distance x from the y axis. The elemental area of the strip is $dA = hdx$.

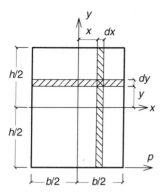

Figure 3.1 Second moment of area of a rectangular section.

Using the second expression in equation 3.1 gives the second moment of area of the section about the y axis as:

$$I_y = \int_{-b/2}^{b/2} x^2 h\,dx = \frac{hb^3}{12} \tag{3.3}$$

Parallel-axis theorem: to determine the second moment of area of a cross-section about an axis which does not pass through the centroid of the section, the parallel-axis theorem may be used. Referring to Figure 3.1, the theorem may be stated in the following manner for the determination of the second moment of area of the section about the axis, p:

$$I_p = I_x + Ad^2 \tag{3.4}$$

where I_x is the second moment of area of the section about the x axis; d is the perpendicular distance between the x axis and the p axis and A is the area of the section. Equation 3.4 states that the second moment of area of the cross-section with respect to the p axis is equal to the sum of the second moment of area about a parallel axis through the centroid and the product of the area and the square of the perpendicular distance between the two axes.

For example, the second moment area of the section in Figure 3.1 about the p axis can be determined using equation 3.1 with y being measured from the p axis and the range of the integration being between 0 and h, i.e.:

$$I_p = \int_0^h y^2 b\,dy = \frac{bh^3}{3} \tag{3.5}$$

Alternatively the parallel axis theorem embodied in equation 3.4 can be used to determine the same second moment of area about the p axis:

$$I_p = I_x + Ad^2 = \frac{bh^3}{12} + bh\left(\frac{h}{2}\right)^2 = \frac{bh^3}{3} \tag{3.6}$$

Relationship between deflection and second moment of area of a beam: the second moment of area of a cross-section is required to calculate the deflection of a beam. Consider a simply supported uniform beam with a span of L, a modulus of elasticity E and a second moment of area I which carries a uniformly distributed load q. The deflection at the mid-span of the beam is:

$$\Delta = \frac{5qL^4}{384EI} \qquad (3.7)$$

Equation 3.7 indicates that the deflection is proportional to the inverse of I. Therefore to reduce the deflection of the beam it is desirable to have the largest I possible for a given amount of material.

Relationship between normal stress and second moment of area of a beam: when a beam is loaded in bending, its longitudinal axis deforms into a curve causing stresses that are normal to the cross-section of the beam. The normal stress, σ_x, in the cross-section at a distance y from the neutral axis can be expressed as:

$$\sigma_x = \frac{My}{I} \qquad (3.8)$$

Equation 3.8 indicates that the normal stress is proportional to the bending moment M and inversely proportional to the second moment of area I of the cross-section. In addition, the stress varies linearly with the distance y from the neutral axis.

Example 3.1

Figure 3.2 shows three different cross-section shapes, A, B and C, two of which are rectangular sections and one is an I-section. They are made up of three identical rectangular plates of width b and thickness of t. The three sections have the same areas. Determine and compare the second moments of areas of the three sections with respect to the horizontal x axis passing through their centroids. For $b = 3t$, $6t$, $9t$, $12t$ and $15t$ compare the relative values of the second moments of area of the sections.

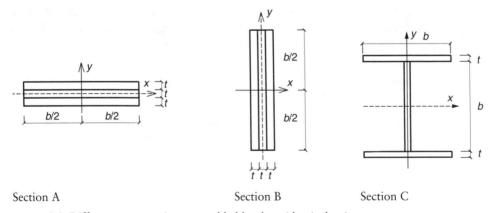

Section A — Section B — Section C
Figure 3.2 Different cross-sections assembled by three identical strips.

Solution

Section A has a width of b and height of $3t$ (Figure 3.2), thus:

$$I_A = \frac{b(3t)^3}{12} = \frac{27bt^3}{12}$$

Section B has a width of $3t$ and height of b (Figure 3.2), therefore:

$$I_B = \frac{3tb^3}{12} = \frac{tb^3}{4}$$

Section C has an I-shape (Figure 3.2). Its second moment of area can be calculated using the parallel axis theorem stated in equation 3.4:

$$I_C = \frac{tb^3}{12} + 2\frac{bt^3}{12} + 2bt\left(\frac{b}{2}+\frac{t}{2}\right)^2 = \frac{7tb^3}{12} + \frac{2bt^3}{3} + b^2t^2$$

The relative values of second moments of area, about the horizontal x axis, with respect to the second moment of area of section A are:

$$\frac{I_B}{I_A} = \frac{3tb^3}{12}\frac{12}{27bt^3} = \frac{b^2}{9t^2} \text{ and}$$

$$\frac{I_C}{I_A} = \left(\frac{7tb^3}{12} + \frac{8bt^3}{12} + \frac{12b^2t^2}{12}\right)\frac{12}{27bt^3} = \frac{7b^2}{27t^2} + \frac{8}{27} + \frac{12b}{27t}$$

The relative I values are given in Table 3.1 for different ratios of b to t.

Table 3.1 Comparison of the relative I values of the three cross-sections

	$b = 3t$	$b = 6t$	$b = 9t$	$b = 12t$	$B = 15t$
Section A	1	1	1	1	1
Section B	1	4	9	16	25
Section C	4	12	25	43	65

It can be observed from Table 3.1 that:

- When $b = 3t$, sections A and B have the same width and height and also have the same second moments of area.
- The ratio of the second moments of area of the section B to section A increases quadratically with the ratio of b/t.
- Section C (the I-shaped section) has the largest second moment of area. When $b = 9t$, its second moment of area is 25 times that of section A and almost 2.8 times that of section B, demonstrating the beneficial use of material in I-sectioned steel members which are extensively used in buildings.
- As demonstrated by section C, the farther the areas of the material in a section are from its centroid, the larger will be the second moment of area.

32 Statics

Example 3.2

Figure 3.3 shows two cross-sections which have the same area. One is a flat section with width 2a and thickness t where $2a \gg t$ and the other is in a folded V-form with its dimensions shown in Figure 3.3b. Calculate the second moments of areas of the two cross-sections about their neutral axes.

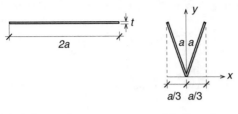

(a) A flat section   (b) V-section

Figure 3.3 Two sections with the same area.

Solution

For the flat form:

$$I_f = \frac{(2a)t^3}{12} = \frac{at^3}{6}$$

For the V-form: the right half of the V-shape can be described by $y = \sqrt{8}x$ (Figure 3.3); the neutral axis can be determined using equation 2.6:

$$c = \frac{2\int_0^{a/3} y\sqrt{1+(y')^2}\,tdx}{2\int_0^{a/3} \sqrt{1+(y')^2}\,tdx} = \frac{2\sqrt{2}a^2t/3}{2at} = \frac{\sqrt{2}}{3}a$$

the second moment of area of the V-section with respect to the x axis is:

$$I = 2\int_0^{a/3} y^2\sqrt{1+(y')^2}\,tdx = \frac{16}{27}a^3t$$

using the parallel axis theorem (equation 3.4), the second moment of area of the V-section about its neutral axis is:

$$I_v = I - Ac^2 = \frac{16a^3t}{27} - 2at\left(\frac{\sqrt{2}a}{3}\right)^2 = \frac{4a^3t}{27}$$

The ratio of the second moments of areas of the V-section to the flat section is:

$$\frac{I_v}{I_f} = \frac{4a^3t}{27} \cdot \frac{6}{at^3} = \frac{8a^2}{9t^2}$$

When $a = 30$ mm and $t = 0.4$ mm, the ratio becomes 5000! This will be demonstrated in section 3.3.2.

Further details can be seen in [3.1].

3.3 Model demonstrations

3.3.1 Two rectangular beams and an I-section beam

This demonstration shows *how the shapes of cross-sections affect the stiffness of a beam*.

The three beams shown in Figure 3.4 can be made using plastic strips, each with the same amount of material, e.g. three 1 mm thick by 15 mm wide strips. Figure 3.4a shows the section of the first beam in which the three strips are glued together to form a section 15 mm wide and 3 mm deep. The beam shown in Figure 3.4b is the same as the first beam but its section is turned through 90 degrees. In the beam shown in Figure 3.4c the three 1 mm thick strips are arranged and glued together to form an I-section. The second moments of area of the three sections about their horizontal neutral axes are in the ratios: $1:25:65$ (Table 3.1). The stiffnesses of the three beams can be demonstrated and felt through simple experiments as follows:

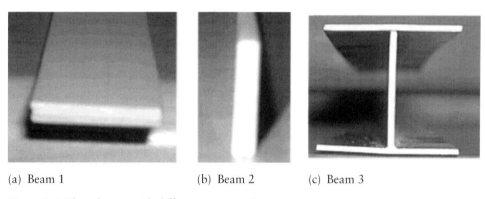

(a) Beam 1   (b) Beam 2   (c) Beam 3

Figure 3.4 Three beams with different cross-sections.

1. Support Beam 1 at its two ends, as shown in Figure 3.5a, and press down at the mid-span of the beam. Notice and feel that the beam sustains a relatively large deflection under the applied load. This beam has a small value of second moment of area because the material of the cross-section is close to its neutral axis.
2. Replace Beam 1 with Beam 2. Press down at the mid-span of the beam whilst also supporting one end of the beam to prevent its tendency to twist (Figure 3.5b). Note that Beam 2 deflects much less than Beam 1 and that its stiffness feels noticeably larger. The material of the cross-section in Beam 2 has been distributed farther away from its neutral axis than was the case for Beam 1, significantly increasing its second moment of area.
3. Replace Beam 2 with Beam 3 and again press down at the mid-span as shown in Figure 3.5c. The beam is stable and feels even stiffer than the rectangular beam

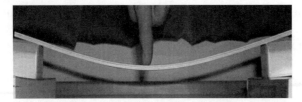

(a) A rectangular section beam with the larger dimension horizontal

(b) A rectangular section with the larger dimension vertical

(c) I-section

Figure 3.5 The deflections of three beams with the same cross-section area but different second moments of area.

shown in Figure 3.5b. The increased stiffness is due to the fact that two-thirds of the material of the cross-section is placed in the flanges, i.e. as far away as possible from the neutral axis of the section.

The simple demonstration shows that the I-section Beam 3 is significantly stiffer than the other two beams and that Beam 2 is much stiffer than Beam 1, although all use the same amount of material.

3.3.2 Lifting a book using a bookmark

This demonstration shows *again how the shapes of cross-sections affect the stiffness of a member*.

Figure 3.6a shows two identical bookmarks which are strips of card with length 210 mm, width 6 mm and approximate thickness 0.4 mm. The lower bookmark in Figure 3.6a is in its original flat form and the upper one is folded and secured with rubber bands to form the V-shape as shown.

If one tries to lift a book using the bookmark in its flat form, the bookmark bends and changes its shape significantly and the book cannot be lifted. This is because the bookmark is very thin and has a small second moment of area around the axis

Effect of different cross-sections 35

passing through the mid thickness of the bookmark, parallel to its width, resulting in a stiffness which is too small to resist the loading.

It is, however, easy to lift the book using the bookmark in its folded form as shown in Figure 3.6b. This is because the folded bookmark has a much larger second moment area since the material is distributed further away from its neutral axis than in the flat form, providing sufficient stiffness to resist the deformation induced by the book. As calculated in example 3.2, the member with the V-shaped section has the stiffness about 5000 times that of the member of the flat section.

(a) Two bookmarks in different shapes   (b) Lifting a book using a folded bookmark

Figure 3.6 Effect of shape of bookmarks.

3.4 Practical examples

3.4.1 A steel-framed building

I-section members are the most commonly used structural elements in steel building frames. Figure 3.7 shows a typical steel-framed building using I-shaped sections for both beams and columns.

Figure 3.7 A steel-framed building.

36 *Statics*

3.4.2 A railway bridge

Figure 3.8 shows a railway bridge. Both the longitudinal and transverse beams are steel I-sections and are used to support the precast concrete slabs of the deck.

Figure 3.8 A railway bridge.

3.4.3 I-section members with holes (cellular beams and columns)

As the material close to the neutral axis of a section does not contribute efficiently to the second moment of area and hence the stiffness of a member, this material can be removed making the member lighter and saving material. It is quite common to see I-section beams and columns with holes along their neutral axes.

Figure 3.9 shows a car showroom in which the external columns support the roof. The external columns are all I-sections with material around their neutral axes

Figure 3.9 Cellular columns.

Figure 3.10 Cellular beams.

removed to create a lighter structure of more elegant appearance. Cellular columns are most effective in cases where axial loads are small. Figure 3.10 shows cellular beams used in an airport terminal.

Reference

3.1 Gere, J. M. (2004) *Mechanics of Materials*, Belmont: Thomson Books/Cole.

4 Bending

4.1 Definitions and concepts

For a beam subjected to bending:

- Elongation occurs on one surface and shortening on the opposite surface of the beam. There is a (neutral) plane through the beam which does not change in length during bending.
- Plane cross-sections of the beam remain plane and perpendicular to the neutral axis of the beam.
- Any deformation of a cross-section of the beam within its own plane is neglected.
- The normal stress on a cross-section of the beam is distributed linearly with the maximum normal stresses occurring on surfaces farthest from the neutral plane.

4.2 Theoretical background

Beams: normally the length of a beam is significantly greater than the dimensions of its cross-section. Many beams are straight and have a constant cross-sectional area.

Shear force and bending moment diagrams: when loaded by transverse loads, a beam develops internal shear forces and bending moments that vary from position to position along the length of the beam. To design beams, it is necessary to know the variations of the shear force and the bending moment and the maximum values of these quantities. The shear forces and bending moments can be expressed as functions of their position x along the length of the beam, which can be plotted along the length of the beam to produce shear force and bending moment diagrams.

Relationships between shear force (V), bending moment (M) and loading (q) can be represented by the following equations:

$$\frac{dV}{dx} = -q(x) \tag{4.1}$$

$$\frac{dM}{dx} = V \tag{4.2}$$

These two equations can be obtained from the free-body diagram for a segment, dx, of a loaded beam (as illustrated in Figure 4.1), making use of vertical force and moment equilibrium. Equations 4.1 and 4.2 are particularly useful for drawing and checking shear force and bending moment diagrams of a beam as they state respectively:

- the slope of the shear force diagram at any point along the length of the beam is equal to minus the intensity of the distributed load;
- the slope of the bending moment diagram at any point along the length of the beam is equal to the shear force at the point.

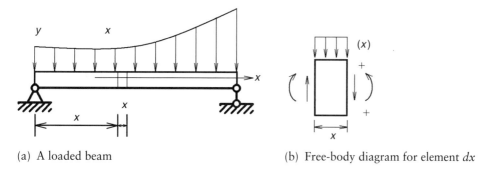

(a) A loaded beam (b) Free-body diagram for element dx

Figure 4.1 Diagrams for deriving equations 4.1 and 4.2.

Example 4.1

Draw the shear force and bending moment diagrams for a simply supported beam subject to a uniformly distributed load q as shown in Figure 4.2a and determine the maximum bending moment.

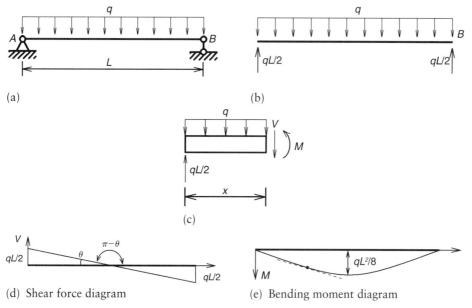

(d) Shear force diagram (e) Bending moment diagram

Figure 4.2 A simply supported beam.

Solution

Taking the free-body diagram of the beam (Figure 4.2b) and using equation 1.1, it is possible to determine the reaction forces from the two supports as indicated in Figure 4.2b. In order to determine the shear force (V) and the bending moment (M) at a cross-section, the equilibrium of a segment of the beam to the left-hand side of the section x as shown in Figure 4.2c can be considered. Applying the two equations of equilibrium gives:

$$\Sigma F_y = 0 \quad \frac{qL}{2} - qx - V = 0 \quad V = q\left(\frac{L}{2} - x\right) \tag{4.3}$$

$$\Sigma M = 0 \quad -\frac{qL}{2}x + qx\frac{x}{2} + M = 0 \quad M = \frac{q}{2}(Lx - x^2) \tag{4.4}$$

The shear force and bending moment diagrams for the beam can be obtained by plotting equations 4.3 and 4.4 as shown in Figures 4.2d and 4.2e. Equations 4.1 and 4.2 can now be used to check the correctness of the shear force and bending moment diagrams. As the distributed load is a constant along the length of the beam (Figure 4.2a), the slope of the shear force diagram is $-[qL/2 - (-qL/2)]L = -q$. According to equation 4.2 and the shear force diagram in Figure 4.2d, the shear force changes from positive to negative linearly so the bending diagram should be a curve. The slope of the bending moment diagram, i.e. the shear force, decreases continuously from $qL/2$ at $x = 0$ to zero at $x = L/2$, then further decreases from zero to $-qL/2$ at $x = L$. The maximum moment occurs at the centre of the beam where the shear force is zero. Thus we have from equation 4.4:

$$M_C = \frac{q}{2}\left[L \cdot \left(\frac{L}{2}\right) - \left(\frac{L}{2}\right)^2\right] = \frac{qL^2}{8} = 0.125qL^2$$

Equation 4.4 also shows that the shape of the bending moment diagram is a parabola.

In order to achieve a safe and economical design, designers often seek ways to reduce the maximum bending moment and the associated stress by reducing spans or creating negative bending moments at supports to offset part of the positive bending moment.

Example 4.2

If the two supports of the beam in Figure 4.2a move inwards symmetrically by distances of μL, the beam becomes an overhanging beam with its ends freely extending over the supports as shown in Figure 4.3a. Determine the value of μ at which the maximum negative and positive bending moments in the beam are the same and determine the corresponding bending moment.

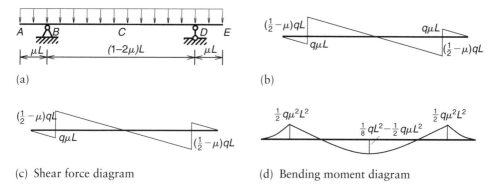

(c) Shear force diagram (d) Bending moment diagram

Figure 4.3 A simply supported beam with overhangs.

Solution

Figure 4.3b shows the free-body diagram for the beam. The bending moments at supports B and D, and at mid-span C can be shown to be:

$$M_B = M_D = -\frac{1}{2}q\mu^2 L^2$$

$$M_C = \frac{1}{2}qL\left(\frac{1}{2} - \mu\right)L - \frac{1}{2}q\left(\frac{L}{2}\right)^2 = \frac{1}{8}qL^2 - \frac{1}{2}q\mu L^2$$

Figure 4.3c and Figure 4.3d show the shear force and bending moment diagrams for the beam. When the magnitudes of M_B and M_C are the same, it leads to:

$$\frac{1}{2}q\mu^2 L^2 = \frac{1}{8}qL^2 - \frac{1}{2}q\mu L^2 \quad \text{or} \quad 4\mu^2 + 4\mu - 1 = 0$$

The solution of this quadratic equation is $\mu = 0.207$. Substituting $\mu = 0.207$ into the expression for M_B or M_C gives:

$$M_B = -\frac{1}{2}q(0.207)^2 L^2 = -0.0214qL^2$$

$$M_C = \frac{1}{8}qL^2 - \frac{0.207}{2}qL^2 = 0.0214qL^2$$

Comparing the maximum bending moments in the simply supported beam ($0.125qL^2$) in Figure 4.2a and the overhanging beam ($0.0214qL^2$) in Figure 4.3a, the former is nearly six times the latter. The reasons for this are:

- the reduced distance between the two supports. Bending moment is proportional to the span squared and the shortened span would effectively reduce the bending moment;

42   Statics

- the negative bending moments over the supports due to the use of the overhangs compensate part of the positive bending moment.

In engineering practice $\mu = 0.2$ is used instead of the exact solution of $\mu = 0.207$ for a simply supported overhanging beam.

Further theoretical background and examples, including the calculation of stresses in beams, can be found in many textbooks, including [4.1, 4.2 and 4.3].

4.3 Model demonstrations

4.3.1 Assumptions in beam bending

This demonstration examines *some of the basic assumptions used in the theory of beam bending*.

A symmetric sponge beam model is made which can be bent and twisted easily (Figure 4.4a). Horizontal lines on the two vertical sides of the beam are drawn at mid-depth, indicating the neutral plane and vertical lines at equal intervals along the length of the sponge are made indicating the different cross-sections of the beam.

Bending the beam as shown in Figure 4.4b, it can be observed that:

- all of the vertical lines, which indicate what is happening to the cross-sections of the beam, remain straight;
- the angles between the vertical lines and the centroidal line (neutral axis) remain at 90 degrees;
- the upper surface of the beam extends and the bottom surface shortens;
- the length of the centroidal (neutral) axis of the beam does not change.

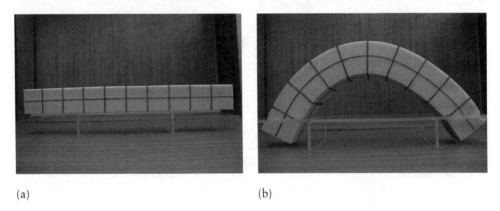

(a)   (b)

Figure 4.4 Examination of beam-bending assumptions.

Bending   43

4.4 Practical examples

4.4.1 Profiles of girders

Large curtain or window walls are often seen at airport terminals. Figure 4.5 shows three such walls. These large window walls are supported by a series of plane girders. The wind loads applied on the window walls are transmitted to the girders and through the girders to their supports.

The girders act like vertical simply supported 'beams'. The bending moments in the 'beams' induced by wind loads are maxima at or close to their centres and minima at their ends. If the wind load is uniformly distributed, the diagrams of bending moments along the 'beams' will be parabolas. Thus it is reasonable to design the girders to have their largest depths at their centres and their smallest depths at their ends with the profiles of the girders being parabolas. The girders shown in Figure 4.5c reflect this and appear more elegant than those shown in Figures 4.5a and 4.5b.

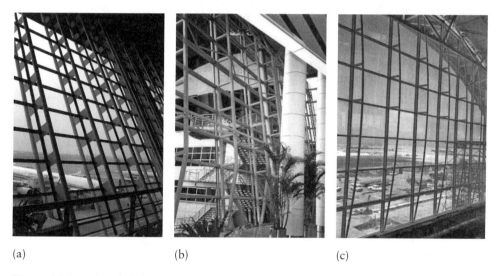

(a)   (b)   (c)

Figure 4.5 A series of girders supporting windows at airport terminals.

4.4.2 Reducing bending moments using overhangs

Figure 4.6 shows a steel-framed multi-storey carpark where cellular beams are used. The vertical loads from floors are transmitted to the beams and then from the beams through bending to the supporting columns. Overhangs are used in the structure which can reduce the bending moments and deflections of the beams. Examining the first overhang, two steel wires are placed to link the free end of the overhang and the concrete support. A downward force on the free end of the overhang is provided by tensions induced in the pair of steel wires. This force will generate a negative bending moment over the column support which will partly offset the positive moments induced in the beam by applied loading.

44 Statics

Figure 4.6 Overhangs used to reduce bending moments and deflections (courtesy of Mr J. Calverley).

4.4.3 Failure due to bending

Figure 4.7a shows a bench which consists of wooden strips, serving as seating and backing, and a pair of concrete frames supporting the seating and backing. Figure 4.7b shows cracks at the end of the cantilever of one of the concrete frames. The two main cracks in the upper part of the cantilever (Figure 4.7b) are a consequence of the normal stresses induced by bending exceeding the limit stress of concrete.

(a) A bench   (b) Cracks at the end of the cantilever

Figure 4.7 A bench with cracks.

4.4.4 Deformation of a staple due to bending

Staplers are common sights in offices. Figure 4.8 shows a typical stapler which creates closed staples through a process of bending and plastic deformation. In this case it can be seen that the two symmetric indents or grooves in the front part of the base of the stapler have bowed shapes.

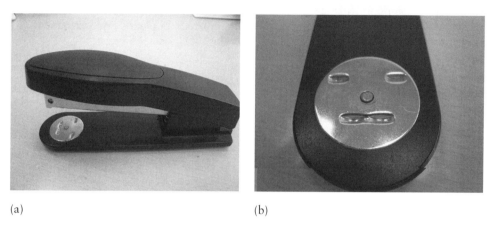

Figure 4.8 A typical stapler.

Figure 4.9 shows the process of the deformation of a staple by the stapler. Initially a staple looks like a frame structure without supports at the two ends of the two vertical elements. When the stapler is depressed, distributed loads are applied on the horizontal element (the middle part of the frame) and these loads push the staple through the multiple sheets of paper being joined to meet the metal base of

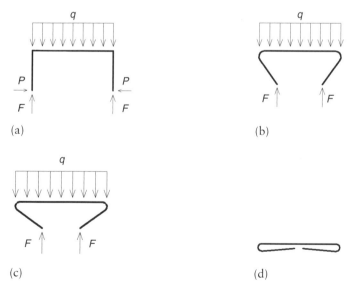

Figure 4.9 The process of the deformation of a staple.

the stapler. When the staple reaches and enters the grooves in the base, both vertical and lateral forces are applied to the ends of the staple as shown in Figure 4.9a. The lateral forces, which are induced due to the shape of the grooves, make the two vertical elements of the staple bend and move inwards. When the staple reaches the position shown in Figure 4.9b, the ends of the staple leave the grooves and no lateral forces act. In this position the vertical forces are sufficient to continue to bend the two vertical elements further (Figure 4.9c). Finally, the staple reaches the shape shown in Figure 4.9d. The large deformations induced are plastic and permanent.

References

4.1 Hibbeler, R. C. (2005) *Mechanics of Materials*, Sixth Edition, Singapore: Prentice-Hall Inc.
4.2 Williams, M. S. and Todd, J. D. (2000) *Structures – Theory and Analysis*, London: Macmillan Press Ltd.
4.3 Gere, J. M. (2004) *Mechanics of Materials*, Belmont: Thomson Books/Cole.

5 Shear and torsion

5.1 Definitions and concepts

Shear: a force (or stress) which tends to slide the material on one side of a surface relative to the material on the other side of the surface in directions parallel to the surface is termed a shear force (or shear stress).

Torsion: a moment that is applied about the longitudinal axis of a member is called a torque which tends to twist the member about this axis and is said to cause torsion of the member.

For a circular shaft or a closed circular section member subjected to torsion:

- plane circular cross-sections remain plane and the cross-sections at the ends of the member remain flat;
- the length and the radius of the member remain unchanged;
- plane circular cross-sections remain perpendicular to the longitudinal axis.

For a non-circular section member or an open section member subjected to torsion:

- plane cross-sections of the member do not remain plane and the cross-sections distort in a manner which is called warping. In other words, the fibres in the longitudinal direction deform unequally.

5.2 Theoretical background

5.2.1 Shear stresses due to bending

When a beam bends, it is also subjected to shear forces as shown in section 4.2, which cause shear stresses in the beam. Considering a vertical cross-section of a loaded, uniform, rectangular section beam, the shear force and shear stresses in the cross-section are as shown in Figure 5.1a. It is reasonable to assume that:

- shear stresses τ act parallel to the shear force V, i.e. parallel to the vertical cross-section;
- the distribution of the shear stresses is uniform across the width (b) of the beam.

Consider a small element between two adjacent vertical cross-sections and between two planes that are parallel to the neutral surface as shown in Figure 5.1b. The existence of the shear stresses on the top and bottom surfaces of the element is due

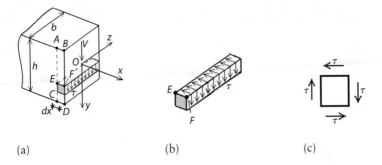

(a) (b) (c)

Figure 5.1 Shear stress on a section [5.1].

to the requirement for equilibrium of the element, which will be demonstrated in section 5.3.2. The shear stresses on the four planes have the same magnitude and have directions such that both stresses point towards or both stresses point away from the line of intersection of the surfaces (Figure 5.1c) [5.1]. Therefore, for determining shear stresses either the vertical or the horizontal planes may be considered.

Figures 5.2a and 5.2b show the element *EFCD* which lies between the adjacent cross-sections *AC* and *BD* separated by a distance *dx* (Figure 5.1a). The shear stresses on *EF* can be evaluated using equilibrium in the direction in which the shear stresses act as [5.1]:

$$\tau = \frac{V \int y \, dA}{bI} = \frac{V A \bar{y}}{bI} \qquad (5.1)$$

where: V is the shear force at the section where the shear stress τ is to be calculated; $\int_A y \, dA$ is the first moment of the area *DFGH* about the neutral axis when τ is to be determined across the surface defined by the line *EF*. For regular sections, the integration can be replaced by the product $A\bar{y}$ in which A is the area of *DFGH* and $\bar{y}$ is the distance between the neutral axis of the cross-section and the centroid of *DFGH*; b is the width of the beam; I is the second moment of area of the cross-section of the beam.

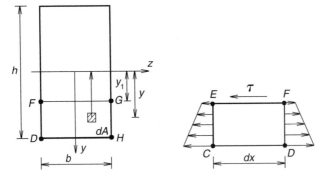

(a) A cross-section   (b) *EFDC* in the plane parallel to the *x*–*y* plane

Figure 5.2 Example 5.1.

Example 5.1

Determine the distribution of shear stress and the maximum shear stress due to a shear force V in the solid rectangular cross-section shown in Figure 5.2 which has a width of b and a depth of h.

Solution

To determine the distribution of the shear stress across the depth of the section, consider a horizontal plane defined by the line FG at a distance of y_1 from the neutral axis and calculate the shear stress along this plane (Figure 5.2b). For the area DFGH, the distance between the centroid of the area and the neutral axis of the section is $\bar{y}$ where:

$$\bar{y} = y_1 + \frac{1}{2}\left(\frac{h}{2} - y_1\right)$$

Thus the first moment of the area about the neutral axis of the section is:

$$A\bar{y} = b\left(\frac{h}{2} - y_1\right) \times \left[y_1 + \frac{1}{2}\left(\frac{h}{2} - y_1\right)\right] = \frac{b}{2}\left[\left(\frac{h}{2}\right)^2 - y_1^2\right]$$

Substituting the expression for $A\bar{y}$ into equation 5.1 gives the shear stress at a depth y_1 from the neutral axis as:

$$\tau = \frac{V}{bI} \times \frac{b}{2}\left[\left(\frac{h}{2}\right)^2 - y_1^2\right] = \frac{V}{2I}\left[\left(\frac{h}{2}\right)^2 - y_1^2\right]$$

This result shows that:

- shear stress varies quadratically along the height of the cross-section;
- the shear stress is zero at the outer fibres of the section where $y_1 = \pm h/2$;
- the maximum shear stress occurs at the neutral axis of the cross-section where $y_1 = 0$, and this maximum stress is:

$$\tau_{max} = \frac{V}{2I}\left(\frac{h}{2}\right)^2 = \frac{V}{2(bh^3/12)}\left(\frac{h}{2}\right)^2 = \frac{3}{2}\frac{V}{bh}$$

The maximum shear stress in a solid rectangular section is 1.5 times the average value of the shear stress across the section.

More detailed information about shear stresses and further examples can be found in [5.1].

5.2.2 Shear stresses due to torsion

Consider a uniform, straight member which is subjected to a constant torque along its length.

If the member has a solid circular section or a closed circular cross-section, the relationship between torque, shear stress and angle of twist is [5.1, 5.2]:

$$\frac{T}{J} = \frac{\tau}{r} = \frac{G\theta}{L} \qquad (5.2)$$

where: T is torque, Nm; J is polar second moment of area, m^4; τ is shear stress, N/m^2 at radius r (m); G is shear modulus, N/m^2; θ is angle of twist, radians, over length L, (m).

Equation 5.2 is derived using the equations of equilibrium, compatibility of deformation and the stress–strain relationship for the material of the member.

For a solid circular member with radius r:

$$J = \frac{\pi r^4}{2} \qquad (5.3)$$

For a hollow circular shaft with inner and outer radii r_1 and r_2 respectively:

$$J = \frac{\pi}{2}(r_2^4 - r_1^4) \qquad (5.4)$$

For a thin-walled member with thickness t and mean radius r:

$$J = 2\pi r^3 t \qquad (5.5)$$

For thin-walled, non-circular sections, the torque–twist relationship in equation 5.2 becomes:

$$\frac{T}{J} = \frac{G\theta}{L} \qquad (5.6)$$

$$J = \frac{4A_e^2}{\int \frac{ds}{t}} \qquad (5.7)$$

where A_e is the area enclosed by the mean perimeter of the section and t is the thickness of the section. When the section has a constant thickness t around its perimeter s, equation 5.7 can be written as:

$$J = \frac{4A_e^2 t}{s} \qquad (5.8)$$

The shear stress across the section is given by:

$$\tau = \frac{T}{2A_e t} \qquad (5.9)$$

For open sections made with thin plates:

$$J = \sum \frac{1}{3} bt^3 \qquad (5.10)$$

where b is the width and t is the thickness of each length of plate which makes up the cross-section.

Example 5.2

There are two members with the same length L, one has a hollow circular section with inner and outer radii r_1 and r_2 respectively (Figure 5.3a) and the other has a similar circular section which has a thin slit cut along its full length (Figure 5.3b). Compare the torsional stiffness of the two members and determine the ratio of the stiffnesses when $r_1 = 12.5\,\text{mm}$ and $r_2 = 40\,\text{mm}$.

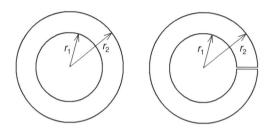

(a) Without a slit   (b) With a slit

Figure 5.3 Hollow circular sections.

Solution

For the closed hollow circular section, equation 5.4 should be used and the stiffness is:

$$\frac{T}{\theta} = \frac{GJ}{L} = \frac{G\pi}{2L}(r_2^4 - r_1^4)$$

The circular section with a split should be treated as a thin rectangular plate using equation 5.10 to determine the stiffness as:

$$\frac{T}{\theta} = \frac{GJ}{L} = \frac{G}{L}\frac{2\pi}{3}\left[r_1 + \frac{(r_2-r_1)}{2}\right](r_2-r_1)^3 = \frac{G\pi}{3L}(r_2+r_1)(r_2-r_1)^3$$

Thus the ratio of the torsional stiffness of the closed section to that of the open section is:

$$\frac{3}{2}\frac{(r_2^4-r_1^4)}{(r_2+r_1)(r_2-r_1)^3}$$

It can be seen that this ratio is independent of the material used for the members of hollow circular sections. Substituting $r_1 = 12.5\,\text{mm}$ and $r_2 = 40\,\text{mm}$ into the above formula gives the ratio to be 3.5. This example will be demonstrated in section 5.3.4.

Example 5.3

Figure 5.4 shows a hollow square section and an I-section each of which has the same cross-sectional area, second moment of area about the x axis and length L. Compare the torsional stiffnesses of the two members when $b = 16$ mm and $t = 1$ mm.

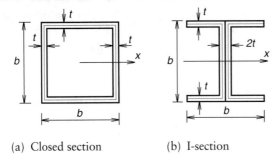

(a) Closed section   (b) I-section

Figure 5.4 Two sections with the same area.

Solution

For the square hollow section: this is a thin-walled non-circular hollow section and equation 5.7 or equation 5.8 should be used to determine the torsional stiffness as:

$$\frac{T}{\theta} = \frac{GJ}{L} = \frac{G}{L}\frac{4A_e^2 t}{s} = \frac{4G}{L}\frac{(b-t)^4 t}{4(b-t)} = \frac{Gt(b-t)^3}{L}$$

For the I-section: this may be considered to be an assembly of three thin plates and equation 5.10 should be used to determine the torsional stiffness as:

$$\frac{T}{\theta} = \frac{GJ}{L} = \frac{G}{L}\sum\frac{bt^3}{3} = \frac{G}{3L}[2bt^3 + (b-2t)(2t)^3] = \frac{2Gt^3(5b-8t)}{3L}$$

Thus the ratio of the torsional stiffnesses of the closed section to the open section is:

$$\frac{3}{2}\frac{(b-t)^3}{t^2(5b-8t)}$$

When $b = 16$ mm and $t = 1$ mm, the ratio is 70. This example will be demonstrated in section 5.3.5.

Examples 5.2 and 5.3 show that a closed section has a much larger torsional stiffness than an open section although the two cross-sections may have the same cross-sectional area and second moment of area. This explains why box girder beams rather than I-section beams are used in practical situations where significant torsional forces are present.

Further theoretical details and examples can be found in [5.1, 5.2, 5.3 and 5.4].

5.3 Model demonstrations

5.3.1 Effect of torsion

This demonstration shows *the effect of shear stress in a non-circular section member induced by torsion.*

Take a length of sponge of rectangular cross-section and mark longitudinal lines down the centre of each face. Then add perpendicular lines at regular intervals along the length of the sponge. Restrain one end of the sponge in a plastic frame as shown in Figure 5.5 and twist the other end. It can be observed that:

- the lines defining the cross-sections of the beam are no longer straight;
- the angles between the horizontal longitudinal lines, which define the neutral axis, and the vertical lines are no longer 90 degrees.

These observations are different from those of the beam in bending, see section 4.3.1.

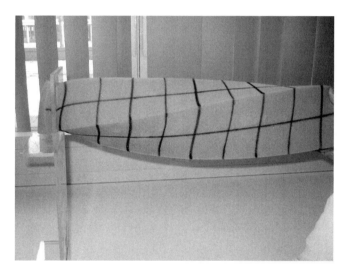

Figure 5.5 Effect of torsion.

5.3.2 Effect of shear stress

This demonstration shows *the existence of shear stress in bending and how shear resistance/stresses between beams/plates/sheets can significantly increase the bending stiffness of a beam.*

Take two identical thick catalogues and drill holes through one of them and then put bolts through the holes and tighten the bolts as shown in Figure 5.6a. Place the two catalogues on a wooden board up against two wood strips, say a quarter of the thickness of a catalogue, which are secured to the board and apply a horizontal force at the top right edge of each of the two catalogues as shown in Figure 5.6b. It is apparent that the thin pages slide over each other in the unbolted catalogue while in the bolted catalogue there is no movement of the pages. This is because the bolts and the friction between the pages provide horizontal shear resistance and prevent the pages sliding between each other.

Support the two catalogues at their ends on two wooden blocks on the board as shown in Figures 5.6c and 5.6d and place the same weight at the mid-span of each of the two catalogues. Figures 5.6c and 5.6d show the bending deflections of the two catalogues with the unbolted catalogue experiencing large deformations while the bolted catalogue experiences only small deformations. The bolts and the friction between the pages provide horizontal shear resistance between the pages of the bolted catalogue and make this act as a single member, a 'thick' beam or plate, whereas in the unbolted catalogue the pages act as a series of very 'thin' beams or plates.

A similar but simpler demonstration of the effect of shear stress can be demonstrated using beams made up of two strips of plastic as shown in Figure 5.7a. For one beam the two plastic strips are loosely bound with elastic bands and for the other beam the two plastic strips are securely held together with four bolts that act as shear connectors and, because of the tension in the tightened bolts, provide compressive forces on the two strips. Suitable sizes for the strips would be 300 mm long, 25 mm wide and 10 mm thick.

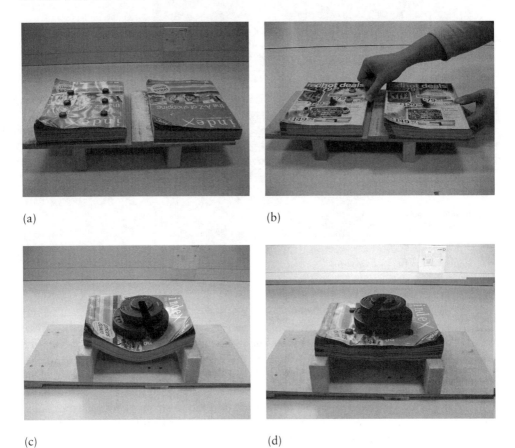

Figure 5.6 Effect of shear stress in catalogues.

By applying similar forces to the ends of the beams one can observe and feel the effect of the shear connection. When bending the beam without shear connectors, one can see the two strips rub against each other and slightly move relative to one another. This becomes noticeable at the ends of the strips as there is little shear resistance between the two strips. When bending the beam with shear connectors, one can feel that the beam is much stiffer than the beam without shear connectors and it is possible to see that there is no relative movement between the strips.

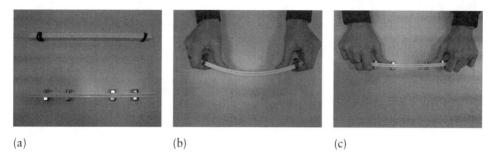

(a)   (b)   (c)

Figure 5.7 Effect of shear resistance in beams.

5.3.3 Effect of shear force

This demonstration shows *how lateral forces can be resisted and transmitted in frame structures through the use of shear elements, such as shear walls.*

Figure 5.8a shows a three-dimensional frame system in which the columns and beams are made of steel springs with wooden joints linking the beams and columns. Applying a horizontal force to the top corner of the frame causes the frame to move in the direction of loading. It can be seen from Figure 5.8a that the angles between beams and columns are no longer 90 degrees as was the case in the unloaded frame.

If a wooden board is fitted into a lower vertical panel of the frame as shown in Figure 5.8b and the same force is applied as was the case in the last demonstration, it is observed that the upper storey experiences horizontal or shear deformations while the lower storey is almost unmoved. The wood panel acts as a shear wall which has a large in-plane stiffness enabling it to transmit horizontal loads in the lower storey directly to the frame supports.

If the wooden board is replaced in a vertical plane in the lower part of one of the end frames as shown in Figure 5.8c and a force is applied at the joint which is one bay away from the wooden board, it is observed that the frame to which the force is applied deforms significantly and the frame where the wooden board is placed has little movement.

If a further wooden board is placed in a horizontal plane, at first-storey level, next to the vertical board as shown in Figure 5.8d and a force is applied at the joint at the corner of the horizontal panel, it can be seen that there is little movement at the point where the force is applied. This is because the horizontal and vertical panels have large stiffnesses in their planes and the shear force is transmitted directly through the wooden boards to the supports.

56 Statics

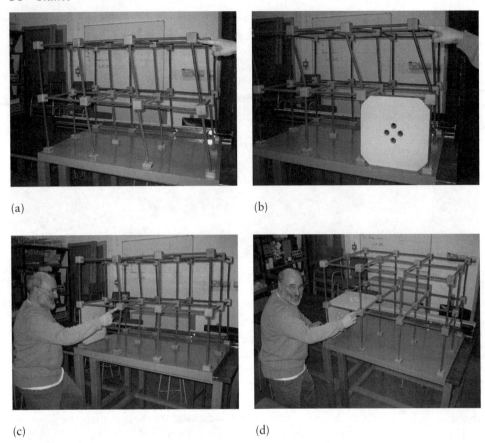

Figure 5.8 Effect of shear walls (the demonstration model was provided by Mr P. Palmer, University of Brighton).

5.3.4 Open and closed sections subject to torsion with warping

This demonstration *shows the difference in the torsional stiffness of two circular members, one with a closed section and the other with an open section where warping can be observed.*

Figure 5.9 shows two foam pipes that are used for insulation and each of the two pipes has a length of 450 mm. The sections have machined slits and in one of the pipes the slit is sealed using tape and glue. One pipe thus effectively has an open circular section and one has a closed circular section. The sections have been analysed in example 5.2 where it was predicted that the torsional stiffness of the closed section was 3.5 times that of the open section. By twisting the two foam pipes it is possible to feel the significant difference in their torsional stiffnesses.

When twisting the two foam pipes with a similar effort it can be seen from Figure 5.9 that:

- it is much easier to twist the open section than the closed section;
- the effect of warping can be observed (Figure 5.9a) at the right-hand end of the pipe which has a slit along its length;

Shear and torsion 57

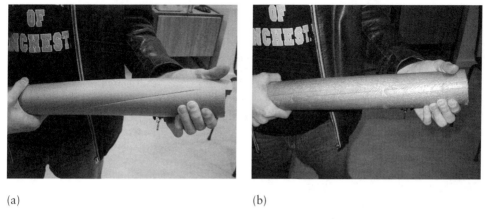

(a)   (b)

Figure 5.9 Open and closed sections subject to torsion with warping.

- there is little warping effect on the pipe with the closed section (Figure 5.9b).

5.3.5 Open and closed sections subject to torsion without warping

This demonstration shows *the difference in the torsional stiffness of two non-circular members, one with a closed section and the other with an open section where warping is restrained.*

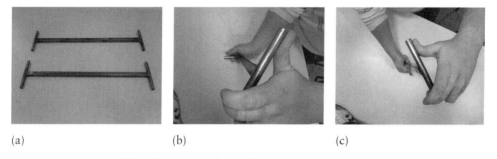

(a)   (b)   (c)

Figure 5.10 Open and closed sections subjected to torsion without warping.

Figure 5.10a shows two 500 mm long steel bars, one with a square hollow section and the other with an I-section which is made by cutting a square hollow section into two halves along its length and welding the resultant channel sections back to back. Handles are welded to the ends of the bars to allow end torques to be easily applied. The sections have been analysed in example 5.3. Due to the addition of the handles, the warping that occurs in open sections as shown in section 5.3.4 is now restrained. In other words, the model with the I-section would be stiffer than that analysed in example 5.3.

By applying torques at the ends of the two bars it is readily felt that the bar with a closed section is much stiffer than the bar which has an open section.

58 *Statics*

5.4 Practical examples

5.4.1 Composite section of a beam

Figure 5.11 shows a number of steel plates or thin beams which are bolted together to form a thick beam. As demonstrated in section 5.3.2, due to the action of the bolts, there are no relative sliding movements between the thin plates/beams when the beam is loaded and due to the shear resistance of the connecting bolts, the thin plates/beams act together as a single member, which is many times stiffer than a member which would result from the plates/beams acting independently.

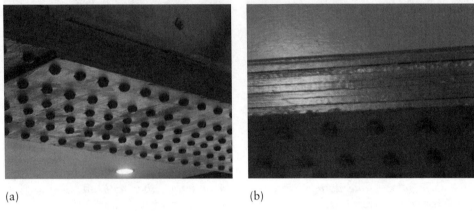

(a)   (b)

Figure 5.11 Composite sections.

5.4.2 Shear walls in a building

Shear walls and bracing members are often present in buildings to provide lateral stiffness and to transmit lateral loads, such as wind loads, to the foundations of the buildings. Figure 5.12 shows one end of a steel-framed building where masonry

Figure 5.12 Shear walls in a steel-framed building.

walls and bracing members are used from the bottom to the top of the building to increase the lateral stiffness of the frame structure.

The dynamic behaviour of this building has been examined, experimentally and numerically, at five distinct construction stages, including the building with and without the walls [5.5]. The walls and bracing members contributed significant stiffness to the building which was reflected in the natural frequencies and their associated mode shapes. The walls and bracing members in the ends of the building increased the fundamental transverse natural frequency from 0.72 Hz for the bare frame structure to 1.95 Hz for the braced structure. The walls in the other two sides of the building were only one-quarter storey height and they increased the fundamental natural frequency from 0.71 Hz to 0.89 Hz.

5.4.3 Opening a drinks bottle

When opening the lid of a common plastic drinks bottle, a torque T applied to the cap is gradually increased until the plastic connectors between the cap and the bottle experience shear failure.

Figure 5.13a shows the connectors which keep the bottle sealed. If there are n connectors, each with an area A, uniformly distributed along the circumference of the cap which has a radius of r and the failure shear stress of the plastic connectors is τ_f, the equation of torsional equilibrium for the cap before being opened is $T = nA\tau_f r$. To open the bottle, the failure shear stress τ_f has to be reached and the design of the cap and its connection to the bottle, in terms of the size of the cap and the number and size of the connectors, needs to be such that this can be achieved with a modest effort. Figure 5.13b shows the shear failure of the connectors.

(a) Connectors in the cap of a drinks bottle   (b) Shear failure of the connectors

Figure 5.13 Opening a drinks bottle.

References

5.1 Gere, J. M. (2004) *Mechanics of Materials*, Belmont: Thomson Books/Cole.
5.2 Benham, P. P., Crawford, R. J. and Armstrong, C. G. (1998) *Mechanics of Engineering Materials*, Harlow: Addison Wesley Longman Ltd.
5.3 Williams, M. S. and Todd, J. D. (2000) *Structures – Theory and Analysis*, London: Macmillan Press Ltd.
5.4 Millais, M. (2005) *Building Structures – From Concepts to Design*, Abingdon: Spon Press.
5.5 Ellis, B. R. and Ji, T. (1996) 'Dynamic testing and numerical modelling of the Cardington steel framed building from construction to completion', *The Structural Engineer*, Vol. 74, No. 11, pp. 186–192.

6 Stress distribution

6.1 Concept

For a given external or internal force, the smaller the area of the member resisting the force, the higher the stress.

6.2 Theoretical background

Normal stress is defined as the intensity of force or the force per unit area acting normal to the area of a member. It is expressed as:

$$\sigma = \lim_{\Delta A \to 0} \frac{\Delta P}{\Delta A} \tag{6.1}$$

where ΔA is an element of area and ΔP is the force acting normal to the area ΔA. When the normal stress tends to 'pull' on the area, it is called *tensile stress* and if it tends to 'push' on the area, it is termed *compressive stress*.

Average normal stress often needs to be considered in engineering practice in such members as truss members, cables and hangers. It is expressed as:

$$\sigma = \frac{P}{A} \tag{6.2}$$

where P is the resultant normal force or external normal force acting on area A. Equation 6.2 assumes:

- after deformation, the area A should still be normal to the force P;
- the material of the member is homogeneous and isotropic. Homogeneous material has the same physical and mechanical properties at different points in its volume and isotropic material has these same properties in all directions at any point in the volume. A typical example of homogeneous and isotropic material is steel.

Equation 6.2 states that *for a given load P, the smaller the area resisting the force, the larger the normal stress*. Further details can be found in [6.1].

62 *Statics*

6.3 Model demonstrations

6.3.1 Balloons on nails

This demonstration shows *the effect of stress distribution*.

Place a balloon on a single nail and hold a thin wooden plate above the balloon and position the balloon as shown in Figure 6.1a. Gradually transfer the weight of the plate onto the balloon. Before the full weight of the plate rests on the balloon, it will burst. This happens because the balloon is in contact with the very small area of the single nail, resulting in very high stress and causing the balloon to burst.

Now place another balloon on a bed of nails instead of a single nail. Put the thin wooden plate on the balloon and add weights gradually onto the plate as shown in Figure 6.1b. It will be seen that the balloon can carry a significant weight before it bursts. It is observed that the shape of the balloon changes but it does not burst. Due to its changed shape, the balloon and hence the weight on the balloon are supported by many nails. As the load is distributed over many nails the stress level caused is not high enough to cause the balloon to burst.

(a) (b)

Figure 6.1 Balloon on nails.

A similar example was observed at a science museum as shown in Figure 6.2. A 6000-nail bed is controlled electronically allowing the nails to move up and down. When the nails move down below the smooth surface of the bed, a young person lies on the bed. Then the 6000 nails move up slowly and uniformly and lift the human body to the position shown in Figure 6.2. If the body has a total uniform mass of 80 kg and about two-thirds of the nails support the body, each loaded nail only carries a force of 0.2 N or a mass of 20 grams. Thus the person is not hurt by the nails.

A similar but simpler demonstration can be conducted following the observation of the nail beds. Figure 6.3a shows 49 plastic cups which are placed upside down and side by side. Place two thin wooden boards on them and invite a person to stand on the boards. The boards spread the weight of the person, 650 N, over the 49 cups, each cup carrying about 13 N, which is less than the 19 N capacity of a cup.

Stress distribution 63

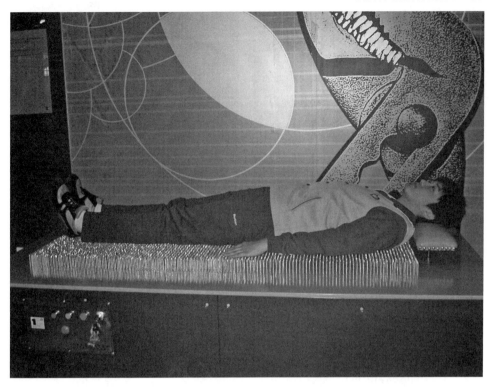

Figure 6.2 A person lying on a nail bed.

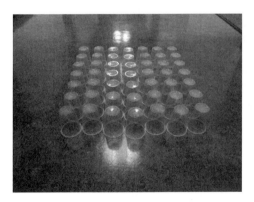

(a) 49 plastic cups placed upside down

(b) A person standing on the cups

Figure 6.3 Uniform force distribution.

6.3.2 Uniform and non-uniform stress distributions

This demonstration shows *the effect of non-uniform stress distribution*.

Place a wooden block on a long sponge and apply a concentrated force at the centre of the block as shown in Figure 6.4a. The sponge under the block deforms uniformly. If the concentrated force is placed at one end of the block as shown in

64 Statics

(a) Uniform stress distribution   (b) Non-uniform stress distribution

Figure 6.4 Uniform and non-uniform stress distributions (the model demonstration was provided by Mr P. Palmer, University of Brighton).

Figure 6.4b, the wooden block rotates with part of the block remaining in contact with the sponge and part separating from the sponge. It can be observed that the sponge deforms non-uniformly, indicating a non-uniform stress distribution in the sponge.

6.4 Practical examples

6.4.1 Flat shoes vs. high-heel shoes

A woman wearing high-heel shoes or flat shoes will exert the same force on a floor but with significantly different stress levels. Figure 6.5 shows a woman weighing 50 kg wearing flat shoes (a) and high-heel shoes (b). The stresses exerted by the shoes can be estimated as follows:

Flat shoes: the contact area of one of the flat shoes (Figure 6.5a) is $12 \times 10^{-3} \, m^2$. Assuming the body weight is uniformly distributed over the contact area, the average stress is:

$$\sigma_{flat} = \frac{50 \times 9.81}{12 \times 10^{-3} \times 2} = 20.4 \, kN/m^2$$

High-heel shoes: assuming that half of the body weight is carried by the high heels (Figure 6.5b); each high heel carries a quarter of the body weight. The area of the high heel is $1.0 \times 10^{-4} \, m^2$. Thus the average stress under each high heel is:

$$\sigma_{heel} = \frac{12.5 \times 9.81}{1 \times 10^{-4}} = 1226 \, kN/m^2$$

This shows that the stress under the heels of the high-heel shoes is about 60 times that under the flat shoes.

A typical adult elephant has a weight of about 5000 kg and the area of each foot is about $0.08 \, m^2$. Thus the average stress exerted by an elephant's foot is:

$$\sigma_{elephant} = \frac{5000 \times 9.81}{0.08 \times 4} = 153.3\,\text{kN/m}^2$$

Elephants have large feet to distribute their body weight. The stress exerted on the ground by the foot of an elephant is much less than that under the high-heel shoes.

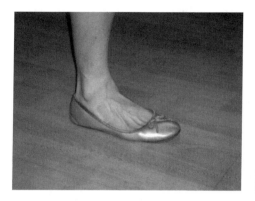

(a) A flat shoe (b) A high-heel shoe

Figure 6.5 Stresses exerted from flat and high-heel shoes (courtesy of Miss C. Patel) [6.2].

6.4.2 The Leaning Tower of Pisa

The tilt of one of the world's most famous towers, the Leaning Tower of Pisa, developed because of uneven stress distribution on the soil supporting the upper structure. The eight-storey tower weighs 14500 metric tonnes and its masonry foundations are 19.6 m in diameter. If the stress was uniformly distributed, this

Figure 6.6 Lead blocks used to reduce the uneven stress distribution on foundation.

would lead to an average stress of 470 kN/m² using equation 6.2. As the underlying ground consists of about 10 m variable soft silty deposits (layer A) and then 40 m very soft and sensitive marine clays (layer B), the Tower shows that the surface of layer B is dish-shaped due to the weight of the Tower above it [6.3]. Thus the uneven stresses on the soft soil under the Tower caused the foundation to settle unevenly, making the Tower lean.

To redress the problem many large blocks of lead were placed on the ground on the side of the Tower where the settlement was least, as shown in Figure 6.6. The new stress in the ground caused the firmer soil on this side of the tower to compress, in turn preventing further leaning of the Tower and returning the Tower back towards the vertical.

References

6.1 Hibbeler, R. C. (2005) *Mechanics of Materials*, Sixth Edition, Singapore: Prentice-Hall Inc.
6.2 Ji, T. and Bell, A. J. (2007) 'Enhancing the understanding of structural concepts – a collection of students' coursework', The University of Manchester.
6.3 Burland, J. B. (2002) 'The Leaning Tower of Pisa', in *The Seventy Architectural Wonders of Our World*, London: Thames & Hudson Ltd.

7 Span and deflection

7.1 Concepts

- For uniformly distributed loads the deflection of a beam is proportional to its span to the power of four and for a concentrated load the deflection is proportional to its span to the power of three.
- To reduce deflections, an increase in the depth of a beam is more effective than an increase its width (for a rectangular cross-section).

7.2 Theoretical background

The relationships between displacement v, slope θ, bending moment M, shear force V and load q for a uniform beam are as follows:

$$\frac{dv}{dx} = \theta \tag{7.1}$$

$$EI\frac{d^2v}{dx^2} = -M \tag{7.2}$$

$$EI\frac{d^3v}{dx^3} = -V \tag{7.3}$$

$$EI\frac{d^4v}{dx^4} = q \tag{7.4}$$

Equation 7.2 indicates that the displacement function v for a loaded beam can be obtained by integrating the bending moment function, M, twice with respect to the coordinate x. Equation 7.4 shows that the displacement function v for a loaded beam can be obtained by integrating the loading function q four times with respect to the coordinate x. When carrying out the integrations, the integration constants can be uniquely determined using the available boundary conditions of the beam.

Sign conventions differ in different textbooks [7.1, 7.2, 7.3 and 7.4]. The sign conventions used in the derivation of the formulae in equations 7.1, 7.2, 7.3 and 7.4 are defined as follows:

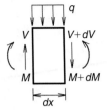

(a) Sign convention for bending moment, shear force and load

(b) Sign convention for displacement and slope

Figure 7.1 Positive sign convention.

The directions of the forces and deflections shown in Figure 7.1 are all positive.

Example 7.1

A simply supported uniform beam subject to uniformly distributed load q is shown in Figure 4.2a. The beam has a length L and constant EI. Use equation 7.2 to determine the displacement of the beam and its maximum deflection.

Solution

The bending moment diagram for the simply supported beam is illustrated in example 4.1 and Figure 4.2e. From equation 4.4, the bending moment at x is:

$$M = \frac{q}{2}(Lx - x^2)$$

Using equation 7.2 gives:

$$EI \frac{d^2v}{dx^2} = -\frac{q}{2}(Lx - x^2)$$

Integrating once with respect to x leads to:

$$EI \frac{dv}{dx} = -\frac{q}{4}Lx^2 + \frac{q}{6}x^3 + C_1$$

From symmetry $dv/dx = 0$ at $x = L/2$. Using this condition the integration constant, C_1, can be determined as $C_1 = qL^3/24$. Thus:

$$\frac{dv}{dx} = -\frac{q}{2EI}\left(\frac{L}{2}x^2 - \frac{1}{3}x^3 - \frac{L^3}{12}\right) \tag{7.5}$$

Span and deflection 69

The slopes at the two ends of the beams can thus be determined and are $qL^3/24$ at $x = 0$ and $-qL^3/24$ at $x = L$. Integrating equation 7.5 gives:

$$v = -\frac{q}{2EI}\left(\frac{L}{6}x^3 - \frac{1}{12}x^4 - \frac{L^3}{12}x\right) + C_2$$

As $v = 0$ at $x = 0$, $C_2 = 0$ and the deflection becomes:

$$v = \frac{q}{12EI}\left(\frac{1}{2}x^4 - Lx^3 + \frac{L^3}{2}x\right) \tag{7.6}$$

The maximum deflection occurs at the centre of the beam, when $x = L/2$. Substituting $x = L/2$ into equation 7.6 gives:

$$v_{max} = \frac{q}{12EI}\left(\frac{1}{32}L^4 - \frac{L^4}{8} + \frac{L^4}{4}\right) = \frac{5qL^4}{384EI} \tag{7.7}$$

Example 7.2

Figure 7.2a shows a uniform beam, fixed at its two ends, carrying a uniformly distributed load q. The beam has a length L and constant EI. Determine the deflections of the beam using equation 7.4.

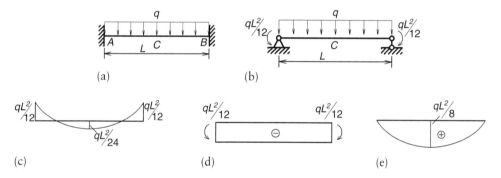

Figure 7.2 A beam with two fixed ends subjected to uniformly distributed loads.

Solution

Equation 7.4 gives:

$$\frac{d^4v}{dx^4} = q$$

Integrating this equation three times and four times leads to:

$$\frac{dv}{dx} = \frac{q}{EI}\left(\frac{x^3}{6} + \frac{C_1}{2}x^2 + C_2x + C_3\right)$$

$$v = \frac{q}{EI}\left(\frac{x^4}{24} + \frac{C_1}{6}x^3 + \frac{C_2}{2}x^2 + C_3x + C_4\right)$$

The four integration constants, C_1 to C_4, can be determined using the boundary conditions of the beam. As $dv/dx = 0$ and $v = 0$ at $x = 0$, it can be shown that $C_3 = 0$ and $C_4 = 0$. The other two integration constants can be determined using two other boundary conditions, $dv/dx = 0$ and $v = 0$ at $x = L$. There is also a symmetry condition available, i.e. $dv/dx = 0$ at $x = L/2$. Using the two slope conditions gives:

$$\frac{qL^3}{6} + \frac{L^2}{2}C_1 + LC_2 = 0$$

$$\frac{qL^3}{48} + \frac{L^2}{8}C_1 + \frac{L}{2}C_2 = 0$$

Solving these two simultaneous equations gives $C_1 = -qL/2$ and $C_2 = qL^2/12$. Thus:

$$v = \frac{q}{12EI}\left(\frac{x^4}{2} - Lx^3 + \frac{L^2}{2}x^2\right) \tag{7.8}$$

Substituting $x = L/2$ into the above equation gives the maximum deflection of the beam at its centre:

$$v_{max} = \frac{qL^4}{384EI} \tag{7.9}$$

The bending moments in the beam can be obtained by using equation 7.2:

$$M = -\frac{q}{2}\left(x^2 - Lx + \frac{L^2}{6}\right) \tag{7.10}$$

Thus the bending moment is $-qL^2/12$ at the two fixed ends and is $qL^2/24$ at the centre of the beam. The bending moment diagram is shown in Figure 7.2c.

It can be seen from Figure 7.2c that the difference in the bending moments at the ends of the beam and the centre of the beam is $qL^2/8$ which is the same as the maximum bending moment in the simply supported beam in example 4.1. Removing the rotational restraints at the ends of the beam in Figure 7.2a and replacing them with two moments as shown in Figure 7.2b, it can be seen that the two beams are equivalent. This means that for the beam with two fixed ends, the bending moment diagram in Figure 7.2c can be interpreted as the summation of the bending diagram due to the end moments, Figure 7.2d, and the bending moment diagram for a simply supported beam carrying the distributed load, Figure 7.2e.

Table 7.1 summarises the maximum bending moments and maximum deflections for a uniform beam with various support conditions carrying either a uniformly distributed load or a concentrated load applied at the most unfavourable position. All the beams have a length L and constant EI.

From Table 7.1 it can be observed that:

- *The maximum deflection of a beam is proportional to its span to the power of four for uniformly distributed loads or to its span to the power of three for a*

Table 7.1 Maximum bending moments and deflections of single-span beams

Boundary conditions and loading conditions	Maximum bending moment	Maximum deflection
Cantilever with point load P at free end B	PL at A	$\dfrac{PL^3}{3EI}$ at B
Simply supported beam with point load P at midspan C	$\dfrac{PL}{4}$ at C	$\dfrac{PL^3}{48EI}$ at C
Fixed-fixed beam with point load P at midspan C	$-\dfrac{PL}{8}$ at A, B; $\dfrac{PL}{8}$ at C	$\dfrac{PL^3}{192EI}$ at C
Cantilever with UDL q	$\dfrac{qL^2}{2}$ at A	$\dfrac{qL^4}{8EI}$ at B
Simply supported beam with UDL q	$\dfrac{qL^2}{8}$ at C	$\dfrac{5qL^4}{384EI}$ at C
Fixed-fixed beam with UDL q	$-\dfrac{qL^2}{12}$ at A, B; $\dfrac{qL^2}{24}$ at C	$\dfrac{qL^4}{384EI}$ at C

  concentrated load. This conclusion is also applicable for other types of distributed loading and for concentrated loads applied at different locations.
- *Larger maximum deflections correspond to situations where larger maximum bending moments occur.* The most common explanation of this observation would lie with the different boundary conditions. Whilst this is true, the explanation can also be stated in terms of the general concept, *the smaller the internal forces, the stiffer the structure*, and this will be explained and demonstrated in detail in Chapters 8 and 9.

7.3 Model demonstrations

7.3.1 Effect of spans

This demonstration shows *the effects of span and second moment of area of a beam on its deflections.*

  A one-metre wooden ruler with cross-section 5 mm by 30 mm is used and a metal

72 Statics

block is attached to one of its two ends. One end of the ruler, with the long side of the cross-section horizontal, is supported to create a cantilever with the concentrated load at its free end.

1. Observe the displacement at the free end of the cantilever with a span of say 350 mm. It can be seen from Figure 7.3a that there is a small deflection at the free end.
2. Double the span to 700 mm as shown in Figure 7.3b and a much larger end displacement is observed. According to the results presented in section 7.2, the end deflection for a span of 700 mm should be eight times that for a span of 350 mm when the effect of the self-weight of the ruler is negligible in comparison to the weight of the metal block.
3. Turn the ruler through 90 degrees about its longitudinal axis as shown in Figure 7.3c and repeat the tests when much reduced end displacement will be seen. The formula in Table 7.1 shows that deflection is proportional to the inverse of the second moment of area which for a rectangular section is given by $I = bh^3/12$. For the current test, the second moment of area of the section about the horizontal axis is 36 times that of the section used in the last test, resulting in maximum deflections of about one thirty-sixth of those in the second test (Figure 7.3b).

(a) (b) (c)

Figure 7.3 Deflections of a cantilever beam subjected to a concentrated load.

7.3.2 Effect of boundary conditions

This demonstration shows *the effects of the boundary conditions, or supports, of a uniform beam on its deflections and shows that fixed boundary conditions produce a stiffer beam than do pinned boundary conditions.*

Figure 7.4 shows the demonstration model which comprises a wooden frame and

(a) Deflections of a simply supported beam   (b) Deflections of a fixed beam

Figure 7.4 Effect of boundary conditions.

two plastic strips with the same length and cross-section. For the fixed beam, a plastic strip is attached securely to the frame with screws and glue at each end and for the simply supported beam a plastic strip is encased at its ends which are free to rotate.

A qualitative demonstration can be quickly conducted. Pressing down at the centres of each of the two beams it is possible to feel qualitatively the difference in the stiffnesses of the two beams. Based on the results in Table 7.1, the fixed beam is four times as stiff as the simply supported beam.

The loads applied and the deflections produced can be quantified. For a particular set of plastic strips, it is found that a concentrated load of 22.3 N produces a measured maximum deflection under the load of 3.5 mm for the beam with the fixed ends whereas for the simply supported beam the maximum deflection is measured to be 13 mm. The ratio of the measured displacements is the same as the theoretical ratio.

7.3.3 The bending moment at one fixed end of a beam

This demonstration shows *how the end moment in a fixed beam can be measured*.

At the fixed end of a beam, both the displacement and the rotation, or slope, are zero. Using the condition that the slope at a fixed end equals zero, a fixed end condition can be created and the moment associated with this condition can be determined.

Figure 7.5a shows a simply supported beam with a supporting frame. A hanger is placed at the centre of the beam so loads can be added. The ends of two vertical arms at the supports are attached to displacement gauges. Readings from the gauges divided by the lengths of the arms are the end rotations of the beam. If weights are placed on two end hangers, they will induce rotations in the opposite directions to those induced by the load applied at the centre of the beam. When the readings in the gauges are reduced to zero, a beam with two fixed ends has been created and the fixed end moments are the products of the weights on the hanger and the horizontal distances between the ends of the hangers and the supports.

Figure 7.5b shows a mass of 5 kg placed on the hanger at the centre of the beam and Figure 7.5c shows the rotation of the beam at the left support and the reading of 2.99 mm. By adding a mass of 3 kg to each of the two end hangers (Figure 7.5d), the gauge at the left end shows a reading of 0.01 mm, indicating that a fixed boundary condition has been created. The associated fixed end moment is $3 \times 9.81 \times 0.125$ (the distance between the end hanger and the support) = 3.68 Nm in this case.

(a) Before loading

(b) Adding a weight at the centre of the beam

74 *Statics*

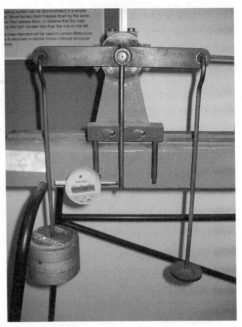

(c) Rotation at the end and the reading (d) Adding loads to remove the rotation

Figure 7.5 Bending moments at the fixed ends of a uniform beam.

7.4 Practical examples

7.4.1 Column supports

Floors which extend outside normal building lines can seldom act as cantilevers due to the effect of the relationship between span, deflection and loading, and thus need additional column supports as shown in Figure 7.6.

Figure 7.6 Column supports.

7.4.2 The phenomenon of prop roots

In rain forests, plants such as Ficus have prop roots. In the humid and shaded conditions, when the large branches reach a certain length, aerial roots grow downwards from the branches. When these aerial roots reach the ground they are similar to stems that support upper branches, forming the unique phenomenon of prop roots.

The prop roots provide the necessary and additional vertical supports to the branches of the tree allowing them to extend their spans further.

Figure 7.7 Prop roots.

7.4.3 Metal props used in structures

The most effective ways to increase the stiffnesses or reduce the deflections of a structure are to reduce spans or to add supports, as shown in the last example obtained from nature.

The Cardiff Millennium Stadium was selected to hold the Eve of the Millennium concert on 31 December 1999. However, the Stadium was designed for sports events rather than for pop concerts. During pop concerts, spectators bounce and jump in time to the music beat and produce dynamic loading on the structure which is larger than the loading due to their static weight. If one of the music beat frequencies occurs at, or is close to, one of the natural frequencies of the cantilever structure, resonance or excessive vibration may occur, which affects both the safety and the serviceability of the structure.

To enable the Eve of the Millennium concert to take place, the cantilever structure of the Cardiff Millennium Stadium had to be reinforced with temporary metal

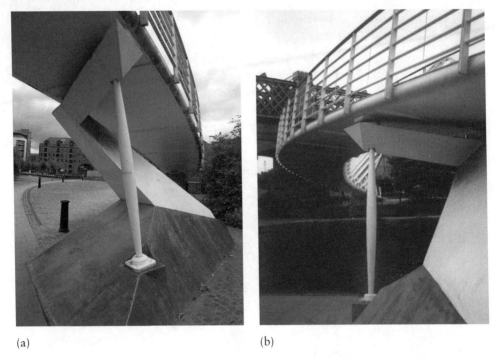

(a) (b)

Figure 7.8 Props used to support a footbridge.

props, which are similar to the prop roots in section 7.4.2. In this way, the spans of the cantilevers were effectively reduced and consequently their stiffnesses and natural frequencies were increased above the range where any unacceptable resonance induced by the spectators was possible.

Figure 7.8 shows a steel prop used to support the deck of a footbridge, which is a critical structural member to the bridge.

References

7.1 Hibbeler, R. C. (2005) *Mechanics of Materials*, Sixth Edition, Singapore: Prentice-Hall Inc.
7.2 Williams, M. S. and Todd, J. D. (2000) *Structures – Theory and Analysis*, London: Macmillan Press Ltd.
7.3 Gere, J. M. (2004) *Mechanics of Materials*, Belmont: Thomson Books/Cole.
7.4 Benham, P. P., Crawford, R. J. and Armstrong, C. G. (1998) *Mechanics of Engineering Materials*, Harlow: Addison Wesley Longman Ltd.

8 Direct force paths

8.1 Definitions, concepts and criteria

Stiffness of a structure is the ability of the structure to resist deformation.

Stiffness of a structure represents the efficiency of the structure to transmit loads on the structure to its supports.

Internal forces in members are induced when they transmit loads from one part to another part of the structure. The internal forces can be tension, compression, shear, torque or bending moment, or a combination of all or some of them.

There are three inter-related concepts relating to the internal forces in a structure:

- the more direct the internal force paths, the stiffer the structure;
- the more uniform the distribution of internal forces, the stiffer the structure;
- the smaller the internal forces, the stiffer the structure.

Following the first concept, five simple criteria can be adopted for arranging bracing members in frame structures to achieve a direct force path leading to a stiffer structure.

- Bracing members should be provided in each storey from the support (base) to the top of the structure.
- Bracing members in different storeys should be directly linked.
- Bracing members should be linked in a straight line where possible.
- Bracing members in the top storey and in the adjacent bays should be directly linked where possible. (Suitable for temporary grandstands and scaffolding structures where the number of bays may be larger than the number of storeys.)
- If extra bracing members are required, they should be arranged following the above four criteria.

8.2 Theoretical background

8.2.1 Introduction

In recent years buildings have become taller, floors wider and bridges longer. It is expected that the trend of increasing heights and spans will continue. How can engineers cope with the ever-increasing heights and spans, and design structures with sufficient stiffness? The basic theory of structures provides the conceptual relationships between span (L), deflection (Δ), stiffness (K_s) and natural frequency (ω) for a single-span beam subject to distributed loads as follows:

$$\Delta = \frac{c_1}{K_s} = c_2 L^4 \tag{8.1}$$

$$\omega = c_3 \sqrt{K_s} = \frac{c_4}{L^2} \tag{8.2}$$

where c_1, c_2, c_3 and c_4 are dimensional coefficients. The two equations state that:

- the deflection of the beam is proportional to its span to the fourth power;
- the fundamental natural frequency of the beam is proportional to the inverse of the span squared;
- both the deflection and fundamental natural frequency are related to the stiffness of the structure.

The limitations on displacements and/or the fundamental natural frequency of a structure specified in building codes actually imply that the structure must possess sufficient stiffness. Adding supports, reducing spans or increasing the sizes of cross-sections of members can effectively increase structural stiffness. However, these measures may not always be possible for practical designs due to aesthetic, structural or service requirements.

The stiffness of a structure is generally understood to be *the ability of the structure to resist deformation*. Structural stiffness describes the capacity of a structure to resist deformations induced by applied loads. Stiffness (K_s) is defined as *the ratio of a force (P) acting on a deformable elastic medium to the resulting displacement (Δ)*, i.e. [8.1]:

$$K_s = \frac{P}{\Delta} \tag{8.3}$$

This definition of stiffness provides a means of calculating or estimating the stiffness of a structure, but does not suggest how to make a structure stiffer. The question of how to design a stiffer structure (the form and pattern of a structure) is a fundamental and practical problem. It may even be a problem that is more challenging than how to analyse the structure.

8.2.2 Concepts for achieving a stiffer structure

8.2.2.1 Definition of stiffness

Consider a structure that consists of s members and n joints, with no limitation on the layout of the structure and the arrangement of members. To evaluate its stiffness at a particular point in a required direction, a unit force should be applied to the point in the direction where the resulting deflection is to be calculated.

Point stiffness is defined as *the inverse of a displacement in the load direction on a node where a unit load is applied.* Thus the point stiffness relates to a unit force which is a function of position and direction. In other words, the point stiffnesses at different positions and in different directions are different.

Define *the stiffness of a structure in a given direction as the smallest value among all point stiffnesses*, i.e.

$$K_s = \min\{k_1, k_2, \ldots, k_j, \ldots, k_n\} \tag{8.4}$$

where K_s is the stiffness, k_j is the point stiffness at the jth node in the given direction, and n is the number of nodes in the structure. Alternatively the stiffness can be expressed as the inverse of the maximum displacement induced by a unit force at the load location in the load direction, i.e.

$$\frac{1}{K_s} = \max\{u_1, u_2, \ldots, u_j, \ldots, u_n\} \tag{8.5}$$

where u_j is the displacement in the load direction of the jth node when a unit force is applied. The node location where the maximum displacement occurs is the *critical point*. The critical points of many structures can be easily identified. For a horizontal cantilever the critical point for a vertical load would be at the free end of the cantilever. For a simply supported rectangular plate the critical point would be at the centre of the plate for a vertical load. For a plane frame supported at its base the critical point would be at the top of the frame for a horizontal load. Thus the stiffness of a structure in a specified direction can be calculated directly by applying the unit load at the critical point in the specified direction.

8.2.2.2 Pin-jointed structures

Consider a pin-jointed structure, such as a truss, containing s bar members and n pinned joints, with a unit load applied at the critical point of the structure. The displacement at the critical point and the internal forces in the members can be obtained by solving the static equilibrium equations and expressed in the following form:

$$1 \times \Delta = \sum_{i=1}^{s} \frac{N_i^2 L_i}{E_i A_i} \tag{8.6}$$

where N_i is the internal force of the ith member induced by a unit load at the critical point; L_i, E_i and A_i ($i = 1, 2, \ldots, s$) are the length, Young's modulus and area of the ith member respectively. Equation 8.6 provides the basis of a standard method for calculating the deflection of pin-jointed structures, and can be found in many textbooks [8.2]. According to the definition given by equation 8.5, the stiffness of the structure is the inverse of the displacement due to a unit load, i.e.

$$K_s = \frac{1}{\sum_{i=1}^{s} \dfrac{N_i^2 L_i}{E_i A_i}} = \frac{1}{\sum_{i=1}^{s} N_i^2 e_i} \tag{8.7}$$

where $e_i = L_i/E_iA_i$ and is known as the flexibility of the ith member. Three concepts embodied in equation 8.6 or equation 8.7 can be explored.

The force N_i in equation 8.7 is a function of structural form and for statically indeterminate structures is also a function of material properties. Therefore, finding the largest stiffness of a pin-jointed structure may be considered as an optimisation problem of structure shape. As equation 8.7 forms an incompletely defined optimisation problem, optimisation techniques may not be applied directly at this stage.

Maximising K_S is achieved by minimising the summation $\sum_{i=1}^{s} N_i^2 e_i$. The characteristics of typical components of the summation are:

1. $e_i > 0$;
2. N_i can be null;
3. $N_i^2 \geq 0$, regardless of whether the member is in tension or compression.

Therefore, to make the summation $\sum_{i=1}^{s} N_i^2 e_i$ as small as possible, three conceptual solutions relating to the internal forces can be developed as follows:

1. as many force components as possible should be zero;
2. no one force component should be significantly larger than the other non-zero forces;
3. the values of all non-zero force components should be as small as possible.

The three conceptual solutions, which are inter-related and not totally compatible, correspond to three structural concepts.

DIRECT FORCE PATH

If many members of a structure subjected to a specific load are in a zero force state, the load is transmitted to the supports of the structure without passing through these members, i.e. the load goes a shorter distance or follows a more direct force path to the supports. This suggests that *shorter or more direct force paths from the load to the structural supports lead to a larger stiffness for a pin-jointed structure*.

UNIFORM FORCE DISTRIBUTION

Consider three sets of data, each consisting of five numbers as shown in Table 8.1. The sums of the three sets of data are the same, but the largest differences between the five numbers in the three sets are different. Consequently, the sums of the square of the three sets are different. The larger the difference of the five numbers in the three sets, the larger the sum of the squares in the example.

Table 8.1 Comparison of three sets of data

Set	Five data	Largest difference	$\sum_{i=1}^{5} a_i$	$\sum_{i=1}^{5} a_i^2$
1	1, 2, 3, 4, 5	4	15	55
2	2, 2, 3, 4, 4	2	15	49
3	3, 3, 3, 3, 3	0	15	45

The comparison between the sums of squares in Table 8.1 shows the effect of the differences between a set of data, which is a simplified case of equation 8.6.

If the largest absolute value of the internal force, $|N_i|$, is not significantly bigger than other absolute values of non-zero forces, it means that the absolute values of the internal forces N_i ($i = 1, 2, \ldots, s$) should be similar. In other words, *more uniformly distributed internal forces result in a bigger stiffness of a pin-jointed structure.*

SMALLER FORCE COMPONENTS

If the values of N_i^2 ($i = 1, 2, \ldots, s$) are small, it means that the force components, either compression or tension, are small. In other words, *smaller internal forces lead to a bigger stiffness of a pin-jointed structure.*

8.2.2.3 Beam types of structure

For a beam type of structure in which bending dominates, an equation, similar to equation 8.6, exists as [8.2]:

$$\Delta = \sum_{i=1}^{s} \int_0^{L_i} \frac{M_i^2(x)}{E_i I_i}\, dx \tag{8.8}$$

where $M_i(x)$, L_i, E_i and I_i are the bending moment, length, Young's modulus and the second moment of area of the cross-section of the ith member respectively. Consider $E_i I_i$ to be constant for the ith member, then the integral $\int_0^{L_i} M_i^2(x)\,dx$ can be expressed by $\overline{M_i^2} L_i$ where $\overline{M_i^2}$ is the median value of $M_i^2(x)$. Thus equation 8.8 becomes:

$$\Delta = \sum_{i=1}^{s} \frac{\overline{M_i^2} L_i}{E_i I_i} \tag{8.9}$$

Equation 8.9 has the same format as equation 8.6 where the numerator contains the square of the internal force. Thus the three concepts derived for pin-jointed structures can also be extended to beam types of structure associated with equation 8.9.

8.2.2.4 Expression of the concepts

As the previous derivation has not been related to any particular material properties, loading conditions or structural form, the three concepts are valid for any bar or beam types of structure and can be used for designing stiffer structures. The three concepts may be summarised in a more concise form as follows [8.3]:

- *The more direct the internal force paths, the stiffer the structure.*
- *The more uniform the distribution of the internal forces, the stiffer the structure.*
- *The smaller the internal forces, the stiffer the structure.*

The concepts are general and valid when equation 8.6 or equation 8.8 is applicable.

82 Statics

8.2.3 Implementation

8.2.3.1 Five criteria

Bracing systems may be used for stabilising structures, transmitting loads and increasing lateral structural stiffness. There are many options to arrange bracing members and there are large numbers of possible bracing patterns, as evidenced in tall buildings, scaffolding structures and temporary grandstands. Five criteria, based on the first concept derived in the last section, have been suggested for arranging bracing members for temporary grandstands. These criteria are also valid for other types of structure, such as tall buildings and scaffolding structures. The five criteria are [8.4]:

- Criterion 1: *bracing members should be provided in each storey from the support (base) to the top of the structure.*
- Criterion 2: *bracing members in different storeys should be directly linked.*
- Criterion 3: *bracing members should be linked in a straight line where possible.*
- Criterion 4: *bracing members in the top storey and in the adjacent bays should be directly linked where possible.*
- Criterion 5: *if extra bracing members are required, they should be arranged following the previous four criteria.*

The first criterion is obvious since the critical point for a multi-storey structure is at the top of the structure and the load at the top must be transmitted to the supports of the structure. If bracing is not arranged over the height of a structure, its efficiency will be significantly reduced. There are a number of ways to achieve the first criterion, but the second and the third criteria suggest a way using a *shorter force path*. The first three criteria mainly concern the bracing arrangements in different storeys of a structure. For some structures, such as temporary grandstands, the number of bays is usually larger than the number of the storeys. To create a shorter force path or more zero force members in such structures, the fourth criterion gives a means for considering the relationship of bracing members across the bays of the structure. The fifth criterion suggests that when extra bracing members are required, usually to reduce the forces of bracing members and distribute the forces more uniformly, they should be arranged using the previous criteria.

Example 8.1

Two four-bay and four-storey plane pinned structures with the same dimensions but with different bracing arrangements are shown in Figure 8.1. All the members are made of the same material and have the same cross-sectional area. The vertical and horizontal members have the same length of one metre. On each structure the same concentrated loads of $0.5\,N$ are applied at the two corner points in the x direction. (If a unit horizontal load is applied on either the top left or top right node, it is difficult to determine the internal forces of all members of the two frames by hand as they are statically indeterminate.) Determine the maximum displacements of the two structures.

Direct force paths 83

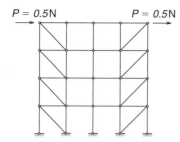

(a) Bracing arrangement following the first criterion (Frame A)

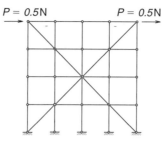

(b) Bracing arrangement following the first three criteria (Frame B)

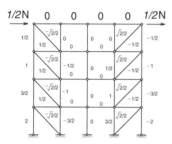

(c) Internal forces (N) in Frame A

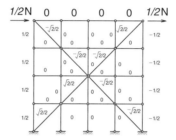

(d) Internal forces (N) in Frame B

Figure 8.1 Two plane frames with different bracing systems.

Solution

The two structures are statically indeterminate. However, they are both symmetric structures subjected to anti-symmetric loads. According to the concept that *a symmetric structure subjected to anti-symmetric loading will result in only anti-symmetric internal forces*, the internal forces in the members of the two frames can be directly calculated using the equilibrium conditions at the pinned joints using the left half of the frames. For example, the internal forces in the horizontal bars in the second and third bays of frame A must be zero as the forces in the two bays must be anti-symmetric and must be in equilibrium at the nodes on the central column. Thus all the internal forces of the two frames can be easily calculated by hand and are marked directly next to the elements as shown in Figures 8.1c and 8.1d, where the positive values mean the members in tension and the negative values indicate the members in compression.

The internal forces are summarised in Table 8.2. The second row shows the magnitudes of the internal forces; the third row gives the numbers of members that have the same force magnitude and the fourth row shows the product N^2L of the corresponding members. ΣN^2L gives the sum of N^2L that have the same force magnitude.

It can be seen from Table 8.2 that:

- there are more zero force members in Frame B than in Frame A;
- the differences between the magnitudes of the internal forces in Frame B are smaller than those in Frame A;

Statics

Table 8.2 Summary of the internal forces of the two frames

	Frame A						Frame B		
Force magnitudes (N)	0	|1/2|	|√2/2|	|1|	|3/2|	|2|	0	|1/2|	|√2/2|
No. of elements	16	10	8	4	4	2	28	8	8
N^2L (N²m)	0	1/4	$\sqrt{2}/2$	1	9/4	4	0	1/4	$\sqrt{2}/2$
ΣN^2L (N²m)	0	5/2	$4\sqrt{2}$	4	9	8	0	2	$4\sqrt{2}$
$\sum_{i=1}^{44} \dfrac{N_i^2 L}{EA}$ (Nm)*	$\dfrac{23.5+4\sqrt{2}}{EA} = \dfrac{29.16}{EA}$						$\dfrac{2+4\sqrt{2}}{EA} = \dfrac{7.657}{EA}$		

Note
*The unit in equation 8.6 is Nm. As the force on the left side of equation 8.6 is one N, the energy and the displacement have the same value. For the studied case, the displacement at the top left and top right nodes of the two frames are the same. Thus $0.5\Delta + 0.5\Delta = 1 \cdot \Delta$

- the magnitudes of the internal forces in Frame B are smaller than those in Frame A.

As Frame A satisfies the first criterion while Frame B satisfies the first three criteria, according to the concepts given in 8.2.2.4 Frame B should be stiffer than Frame A. The maximum displacements of the two frames induced by the same loading are given at the bottom row in Table 8.2. In other words, the lateral stiffness of Frame B is 3.81 times (29.16/7.657) that of Frame A, although the same amount of material is used in the two frames. This demonstrates the effect of using the concepts and criteria defined earlier. Experimental and physical models of the two frames will be provided to demonstrate the difference between the lateral stiffness of the two frames in section 8.3.

8.2.3.2 Numerical verification

To examine the efficiency of the concepts, consider a pin-jointed plane frame, consisting of four bays and two storeys with six different bracing arrangements as shown in Figure 8.2. All frame members have the same Young's modulus E and cross-sectional area A with EA equal to 1000 N. The vertical and horizontal members have unit lengths (one metre). A concentrated horizontal load of 0.2 Newton is applied to each of the five top nodes of the frames. The lateral stiffness can be calculated as the inverse of the averaged displacement of the top five nodes in the horizontal direction.

The bracing members in the six frames are arranged in such a way that the efficiency of each criterion given in section 8.2.3.1 can be identified. The features of the bracing arrangements can be summarised as follows:

- the bracing members in Frame (a) satisfy the first criterion;
- the bracing members in Frame (b) satisfy the first two criteria;
- the bracing members in Frame (c) satisfy the first three criteria;
- the bracing members in Frame (d) satisfy the first four criteria;
- two more bracing members are added to Frame (c) to form Frame (e), but the added bracing members do not follow the criteria suggested;
- four more bracing members are added to Frame (c) or Frame (d) to form Frame (f), and the arrangement of bracing members in Frame (f) satisfies all of the five criteria.

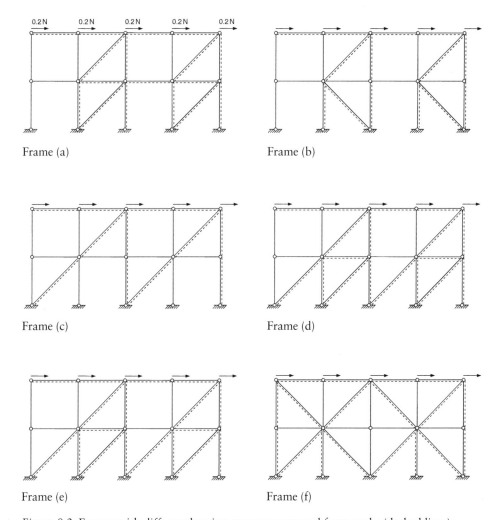

Figure 8.2 Frames with different bracing arrangements and force paths (dashed lines).

Table 8.3 lists the total numbers of members, bracing members and zero-force members, the five largest absolute values of member forces and the average horizontal displacements of the five top nodes of the six frames. The relative stiffnesses of the six frames are also given for comparison.

The force paths, which transmit the loads from the tops to the supports of the frames, are indicated by the dashed lines in Figure 8.2. To emphasise the main force paths, forces of less than 3 per cent of the maximum force in each of the first three frames have been neglected in Figure 8.2.

Following the concepts suggested on the basis of equation 8.7, it can be seen from Table 8.3 and Figure 8.2 that:

- Frame (a) has a conventional form of bracing and the loads at the top are

Table 8.3 A summary of the results for the six frames (Figure 8.2)

Frame	(a)	(b)	(c)	(d)	(e)	(f)
No. of elements	22	22	22	22	24	26
No. of bracing elements	4	4	4	4	6	8
No. of zero-force elements	5	8	10	14	6	8
The absolute values of the five largest element forces (N) (b – bracing members) (v – vertical members)	1.04 (v) 0.96 (v) 0.78 (b) 0.72 (b) 0.69 (b)	1.03 (v) 0.97 (v) 0.74 (b) 0.71 (b) 0.71 (b)	0.75 (b) 0.72 (b) 0.69 (b) 0.67 (b) 0.53 (v)	0.71 (b) 0.71 (b) 0.71 (b) 0.71 (b) 0.40 (v)	0.74 (b) 0.67 (b) 0.64 (v) 0.59 (b) 0.58 (v)	0.40 (b) 0.40 (b) 0.37 (b) 0.37 (b) 0.33 (b)
The average horizontal displacement of the five top nodes (mm)	6.60	6.12	4.12	3.23	3.93	1.69
The relative stiffness	1	1.08	1.60	2.04	1.68	3.91
The ratio of stiffness to the total area of bracing members	1	1.08	1.60	2.04	1.12	1.95

transmitted to the base through the bracing, vertical and horizontal members. There are five members with zero force.
- In Frame (b) the forces in the bracing members in the upper storey are directly transmitted to the bracing and vertical members in the lower storey without passing through the horizontal members that link the bracing members in the two storeys. Thus Frame (b) provides a shorter force path with three more zero-force members and yields a higher stiffness than Frame (a).
- In Frame (c) a more direct force path is created with two vertical members in the lower storey, which have the largest forces in Frame (b), becoming zero-force members. The shorter force path produces an even higher stiffness, as expected.
- To transmit the lateral loads at the top nodes where bracing members are involved, forces in vertical members have to be generated to balance the vertical components of the forces in the bracing members in Frame (c). In Frame (d) two bracing members with symmetric orientation are connected at the same node, with one in compression and the other in tension. The horizontal components of the forces in these bracing members balance the external loads while the vertical components of the forces are self-balancing. Therefore, all vertical members are in a zero-force state and Frame (d) leads to the highest stiffness of Frames (a)–(d).
- Two more members are added to Frame (c) to form Frame (e), but comparison between Frame (d) and Frame (e) indicates that bracing members following the criteria set out can lead to a higher stiffness than more bracing members which do not fully follow the criteria.
- Frame (f) shows the effect of the fifth criterion. Four more bracing members are added to Frame (d) and arranged according to the first three criteria. Now the lateral loads are distributed between more members, creating a smaller and a more uniform force distribution, which results in an even higher stiffness.

It can be seen from Table 8.3 that the structure is stiffer when the internal forces are smaller and more uniformly distributed although the first four criteria are derived based on the concept of direct force paths. These examples are simple and the variation of bracing arrangements is limited, but they demonstrate the efficiency of the concepts and the criteria.

The lateral stiffnesses of the frames are provided by the bracing members. It is interesting to examine the ratio of the relative lateral stiffness to the total area of bracing members. In this way, frame (d) has the highest ratio.

8.2.4 Discussion

Design of a structure needs to consider several aspects and the stiffness requirement is one of them. The concept of direct force paths and the criteria which follow from it may be useful for design of those structures when increasing stiffness is important.

8.2.4.1 Safety, economy and elegance

A useful definition of Structural Engineering has been given in the *Journal of the UK Institution of Structural Engineers* [8.5] as follows:

> *Structural engineering is the science and art of designing and making, with economy and elegance, buildings, bridges, frameworks, and other similar structures so that they can safely resist the forces to which they may be subjected.*

There are three key factors in the statement: *safety, economy* and *elegance*. The discipline of structural engineering allows structures to be produced with satisfactory performance at competitive costs. Elegance, which is not particularly related to safety and economy, should also be considered. However, the beauty of the three structural concepts presented in this chapter lies in integrating safety, economy and elegance of a structure as a whole. This may be demonstrated by the following example.

Figure 8.3a shows the bracing arrangement at the back of part of an actual temporary grandstand, where alternative bays were braced from the bottom to the top and a bracing member is placed on first-storey level in all the other bays. It can be seen that the bracing system satisfies the first criterion and partly satisfies the third criterion (section 8.1). A significant increase in the lateral stiffness of the grandstand can be achieved by using the concept of direct force paths. Without considering the safety, economy and elegance of the structure but following the concept of the direct force paths and the first four criteria, the bracing members can be re-arranged as shown in Figure 8.3b. The calculated lateral stiffnesses of the two frames are summarised in Table 8.4 [8.4].

The comparison shows that the lateral stiffness of the improved structure is higher, being 284 per cent of the stiffness of the original structure. The improved structure is also more economical as the number of bracing members is reduced by 19 per cent. Looking at the appearance of the two frames, one would probably feel that the frame with the improved bracing arrangement is more elegant.

Application of the concepts leads to structures with larger stiffness and smaller and more uniform distributions of internal forces, meeting the requirements of

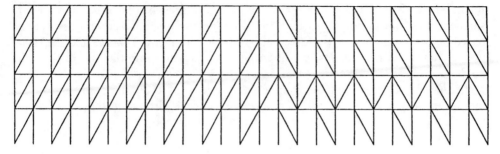

(a) Original bracing system

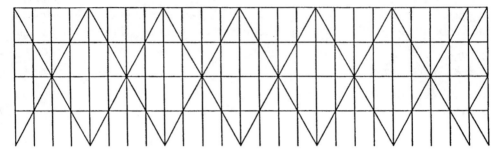

(b) Improved bracing system

Figure 8.3 Bracing arrangements for a temporary grandstand [8.4].

Table 8.4 Comparison of the efficiency of two braced frames (Figure 8.3)

	Lateral stiffness (MN/m)	No. of bracing members used
Original Frame (OF) (Figure 8.3a)	3.16	64
Improved Frame (IF) (Figure 8.3b)	8.96	52
(IF)/(OF)	284%	81%

safety and economy. It is difficult to show that the concepts also lead to elegant designs. However, the examples of the John Hancock Tower, the Bank of China (section 8.4.1) and the Raleigh Arena (section 9.4.1) are all well-known safe, economical and elegant structures in which the concepts presented in section 8.1 were used.

8.2.4.2 Optimum design and conceptual design

The three concepts presented are derived on the basis of making displacements (equation 8.6 or equation 8.9) as small as possible. It is useful to compare the general characteristics of optimum design methods and the use of the concepts. Table 8.5 compares the general characteristics of some optimum design methods and the design methods using the proposed concepts.

Table 8.5 Comparison of design methods

	Design using optimum methods	Design using the concepts
Objective	Seek a maximum or minimum value of a function, such as cost, weight or energy	Seek a stiffer structure rather than the stiffest structure
Constraints	Explicitly applied	Implicitly applied
Solution method	Computer-based mathematical methods	The concepts and derived criteria
Loading	The optimum design depends on loading conditions	The design is independent of loading conditions
The cross-sectional sizes of members	Provided as the solution of the optimum design	To be determined
Design	The optimum design may not be practical	The design is practical
Users	Specialists and researchers	Engineers

Compared to optimum design methods, design based on the concepts presented does not involve an analysis for choosing member cross-sections, does not seek the stiffest structure and is not subjected to explicitly applied constraints. Therefore, design using the concepts becomes simple and many engineers can make direct use of the concepts in their designs. It is useful that the choice of structural form, relating to force paths, and selection of the sizes of cross-sections are conducted separately.

As the concepts are fundamental and they can be applied to a structure globally, designs of bracing arrangement based on the concepts may be more optimal and rational than some designs resulting from optimum design processes. Figure 8.4a shows the optimal topology of a bracing system for a steel building framework with an overall stiffness constraint under multiple lateral loading conditions [8.6]. The solution was obtained by gradually removing the elements with the lowest strain energy from a continuum design, which implied creating a direct force path in an iterative manner. Figure 8.4b shows the design of the bracing system using the concept of direct force paths, i.e. the first three criteria. The design of the bracing system takes only a few minutes. Using the dimensions, cross-sectional sizes and

Table 8.6 Displacements of the two designs

	Frame without bracing members	Frame with bracing suggested in [8.6]	Frame with bracing using the first concept
Horizontal displacement at the top of the frame	630 mm	87.4 mm	78.5 mm
Maximum horizontal displacement and its location	630 mm Top-storey level	87.4 mm Top-floor level	84.1 mm 11th-storey level
Bracing members used	0	100%	67.5%

90 Statics

load conditions given in [8.6], the maximum displacements of the two frames have been calculated and are listed in Table 8.6.

For this example, it can be seen that the design using the first concept is more practical, economical and is stiffer than the design obtained using an optimisation technique.

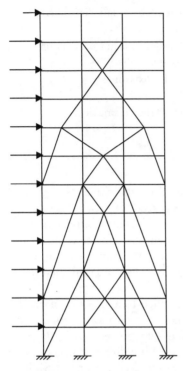
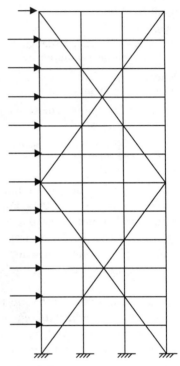

(a) Optimal topology of bracing system [8.6] (b) Bracing system following the concept of direct force path

Figure 8.4 Comparison of two designs of bracing system.

8.3 Model demonstrations

8.3.1 Experimental verification

These simple experiments *verify the concept that the more direct the internal force paths, the stiffer the structure, and the first three corresponding criteria.*

Three aluminium frames were constructed with the same overall dimensions of 1025 mm by 1025 mm. All members of the frames have the same cross-section of 25 mm by 3 mm. The only difference between the three frames is the arrangement of the bracing members as shown in Figure 8.3. It can be seen from Figure 8.5 that:

1   Frame A is traditionally braced with eight members, which satisfies the first criterion.

2   Eight bracing members are again used in Frame B but are arranged to satisfy the first three criteria.
3   A second traditional bracing pattern is used for Frame C with 16 bracing members arranged satisfying the first two criteria.

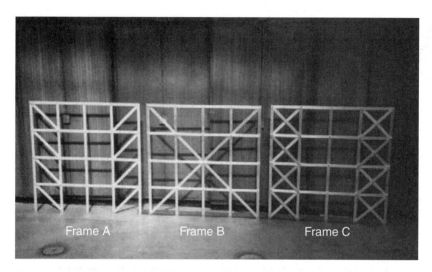

Figure 8.5 Aluminium test frames (Frames A, B and C are placed from left to right).

The three frames were tested using a simple arrangement. The frames were fixed at their supports and a hydraulic jack was used to apply a horizontal force at the top right-hand joint of the frame. A micrometer gauge was used to measure the horizontal displacement at the top left-hand joint of the frame. A lateral restraint system was provided to prevent out-of-plane deformations [8.7].

The horizontal load-deflection characteristics of the three frames are shown in Figure 8.6. It can be seen that the displacements of Frame B, which satisfied the first three criteria, are about one-quarter of those of Frame A for the same load. Frame C

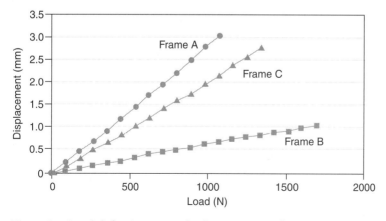

Figure 8.6 Load-deflection curves for frames A, B and C.

92 *Statics*

with eight more members but not satisfying the third criterion is obviously less stiff than Frame B. For example, the displacements corresponding to the load of about 1070 N are 3 mm for Frame A, 0.73 mm for Frame B and 2.2 mm for Frame C respectively. The experiment results for Frame A and Frame B align with the conclusions obtained in example 8.1.

8.3.2 *Direct and zigzag force paths*

This model demonstration allows *one to feel the relative stiffnesses of two similar plastic frames and shows the effect of internal force paths.*

In order to 'feel' the effect of the force paths, two frames were made of plastic, with the same overall dimensions 400 mm by 400 mm and member sizes of 25 mm by 2 mm. See Figure 8.7 [8.7]. The only difference between the two frames is the arrangement of bracing members. The forms of the two frame models are the same as the example calculated in section 8.2.3 and the test frames A and B in section 8.3.1. The relative stiffnesses of the two frames can be felt by pushing a top corner joint of each frame horizontally. The frame on the right side feels much stiffer than the one on the left. In fact, the stiffness of the right frame is about four times that of the left frame. The load applied to the right frame is transmitted to its supports through a direct force path while for the frame on the left, the force path is zigzag.

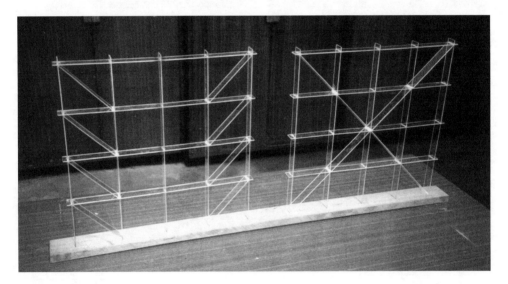

Figure 8.7 Braced frame models showing direct and zigzag force paths.

8.4 Practical examples

8.4.1 *Bracing systems of tall buildings*

The 100-storey 344 m tall building, the John Hancock Center in Chicago, has an exterior-braced frame tube structure. An advance on the steel-framed tube, this design added global cross-bracing to the perimeter frame to increase the stiffness of the structure as shown in Figure 8.8a. Some $15 million was saved on the conventional steelwork by using these huge cross-braces [8.8]. It was regarded as an extremely economical design which achieved the required stiffness to make the building stable. One of the reasons for the success was, as can be seen from Figure 8.8a, that the required lateral stiffness of the structure was achieved by using the cross-braces resulting in direct force paths and smaller internal forces according to the first concept or the first three criteria (sections 8.1 and 8.2). The Bank of China, Hong Kong (Figure 8.8b) also adopts a similar bracing system.

(a) John Hancock Center   (b) Bank of China in Hong Kong

Figure 8.8 Bracing systems used in buildings satisfying the first three criteria.

8.4.2 Bracing systems of scaffolding structures

Scaffolding structures are temporary structures that are an essential part of the construction process. Scaffolding imposes certain design restrictions that can be ignored in the design of other structures. For example, scaffolding structures must be easily assembled and taken apart, and the components should also be relatively light to permit construction workers to handle them. Although scaffolding structures are light and temporary in the majority of cases, their design should be taken seriously. The concept of direct force paths and the five criteria are applicable to scaffolding structures.

8.4.2.1 The collapse of a scaffolding structure

The scaffolding structure shown in Figure 8.9 collapsed in 1993 [8.9], although no specific explanation was given. Using the concept of direct force paths and the

Figure 8.9 Collapse of a scaffolding structure (courtesy of Mr J. Anderson).

understanding gained from the previous examples, the cause of the incident may be explained. It can be seen that in this scaffolding structure no diagonal (bracing) members were provided, i.e. no direct force paths were provided. The scaffolding structure worked as an unbraced frame structure, and the lateral loads, such as wind loads, on the structure were transmitted to its supports through bending of the slender scaffolding members. The structure did not have enough lateral stiffness and collapsed under wind loads only.

8.4.2.2 *Some bracing systems used for scaffolding structures*

For the convenience of erection of the scaffolding structures shown in Figure 8.10, standard units were used. The unit shown in Figure 8.10a consists of two horizontal members, two vertical members and two short bracing members. The unit is useful for transmitting the vertical loads applied to the top horizontal member to the vertical members that support the unit at its two ends. The unit is equivalent to a thick beam in the structure and the scaffolding structure becomes a deep beam and slender column system. The diagonal members used in the structure do not provide the force paths to transmit the lateral loads on the structure from the top to the bottom of the structure and do not follow the basic criteria for arranging bracing members. Therefore it can be seen that the scaffolding structure has a relatively low lateral stiffness based on the first concept of direct force paths.

Bracing members are also provided in the scaffolding structure shown in Figure 8.10b. However, these bracing members are linked in the horizontal direction but not connected from the top to the bottom of the structure, and do not create direct force paths. Therefore, without any calculation, it can be judged that the scaffolding structure possesses a relatively low lateral stiffness.

(a)   (b)

Figure 8.10 Inefficient bracing systems for scaffolding structures.

References

8.1 Parker, S. P. (1997) *Dictionary of Engineering*, Fifth Edition, New York: McGraw-Hill.
8.2 Gere, J. M. (2004) *Mechanics of Materials*, Belmont: Thomson Books/Cole.
8.3 Ji, T. (2003) 'Concepts for designing stiffer structures', *The Structural Engineer*, Vol. 81, No. 21, pp. 36–42.
8.4 Ji, T. and Ellis, B. R. (1997) 'Effective bracing systems for temporary grandstands', *The Structural Engineer*, Vol. 75, No. 6, pp. 95–100.
8.5 *The Structural Engineer*, Vol. 72, No. 3, 1994.
8.6 Lian, Q., Xie, Y. and Steven, G. (2000) 'Optimal topology design of bracing systems for multi-story steel frames', *Journal of Structural Engineering*, ASCE, Vol. 126, No. 7, pp. 823–829.
8.7 Roohi, R. (1998) 'Analysis, testing and model demonstration of efficiency of different bracing arrangements', Investigative Project Report, UMIST.
8.8 Bennett, D. (1995) *Skyscrapers: Form and Function*, New York: Simon & Schuster.
8.9 Anderson, J. (1996) 'Teaching health and safety at university, Proceedings of the Institution of Civil Engineers', *Journal of Civil Engineering*, Vol. 114, No. 2, pp. 98–99.

9 Smaller internal forces

9.1 Concepts

- The more direct the internal force paths, the stiffer the structure.
- The more uniform the distribution of internal forces, the stiffer the structure.
- The smaller the internal forces, the stiffer the structure.

Smaller internal forces in a structure or a stiffer structure can be achieved by:

- providing additional supports to the structure;
- reducing spans of the structure; or
- making a self-balanced system of forces in the structure before the forces are transmitted to the supports of the structure.

The first two measures are obvious. The contents of this chapter are related to the third measure.

Criterion: if members can be added into a structure in a way that offsets some of the effects of the external loads or balances some of the internal forces before the forces are transmitted to the supports of the structure. The internal forces in the structure will be smaller and the structure will be stiffer.

9.2 Theoretical background

9.2.1 Introduction

In Chapter 8 the ways to achieve direct force paths between the load at the critical point of a structure and its supports were discussed leading to a stiffer and more economical design. In this chapter the ways to achieve smaller internal forces in a structure will be considered and this will also lead to a stiffer and more economical design.

Increasing the sizes of the cross-sections of members in a structure will effectively reduce their stress levels but not necessarily the internal forces in the members. This measure usually increases the amount of material used and therefore the weight of the structure. Reducing spans is a very effective way to reduce the magnitudes of internal forces and increase the stiffness of the structure. However, this measure may not be feasible in many practical cases. Following the third concept derived in Chapter 8, if the internal forces can be partly balanced by introducing new structural members, smaller internal forces will be created and the structure will be stiffer. The theoretical background for this concept and for the criterion given in section 9.1 are provided in this section.

Consider a beam type of structure with s members and a concentrated load P applied at the critical point of the structure. The maximum displacement of the structure in the loading direction can be expressed as [9.1, 9.2]:

$$v_1 = \sum_{i=1}^{s} \int_0^{L_i} \frac{M_i^P(x)\overline{M}_i^P(x)}{E_i I_i} dx \tag{9.1}$$

where $M_i^P(x)$ and $\overline{M}_i^P(x)$ are the bending moments of the ith member induced by the load and a unit force applied respectively on the critical point in the direction where the displacement is to be calculated. Thus $M_i^P(x)$ and $\overline{M}_i^P(x)$ have the same form and sign. If a structural member is added into the structure, the internal forces in the structure are consequently changed. The internal forces of the modified structure induced by the load can be expressed as the summation of the internal forces, $M_i^P(x)$, of the original structure and the change, $\Delta M_i(x)$, ($i = 1, 2, ..., s$). Retaining the force effect of the member on the structure but neglecting its effect on the stiffness of the structure (i.e. the change of the stiffness matrix), the displacement at the same location becomes [9.1]:

$$v = \sum_{i=1}^{s} \int_0^{L_i} \frac{M_i^P(x)\overline{M}_i^P(x)}{E_i I_i} dx + \sum_{i=1}^{s} \int_0^{L_i} \frac{\Delta M_i(x)\overline{M}_i^P(x)}{E_i I_i} dx = v_1 + v_2 \tag{9.2}$$

where

$$v_2 = \sum_{i=1}^{s} \int_0^{L_i} \frac{\Delta M_i(x)\overline{M}_i^P(x)}{E_i I_i} dx \tag{9.3}$$

If the member is positioned in the structure so that many of the terms $\Delta M_i(x)$ have the opposite signs to $M_i^P(x)$ and $\overline{M}_i^P(x)$, the second term in equation 9.2 will become negative. Therefore the displacement in equation 9.2 will be smaller than that in equation 9.1. In other words, the function of the added member is to create additional internal forces that have the opposite directions to those induced by the concentrated load P in the original structure. This reduces the magnitudes of the internal forces and yields a stiffer structure. The position of the member can be identified from the deflected shape of the structure and the form of which can be determined by experience or by calculations. This is best illustrated through examples.

9.2.2 A ring and a tied ring

A ring, with radius R, has a rigidity of EI and is subjected to a pair of vertical forces P at points B and D as shown in Figure 9.1a [9.3]. Similar to studying the displacements and internal forces of a straight beam, it may be assumed that the ring experiences small deflections allowing the equilibrium equations to be established using the configurations before deformation. The deformation of the ring is dominated by bending. Due to symmetry, only the right top quarter of the ring needs to be considered for analysis (Figure 9.1b). Also due to the double symmetry of the loaded

98 Statics

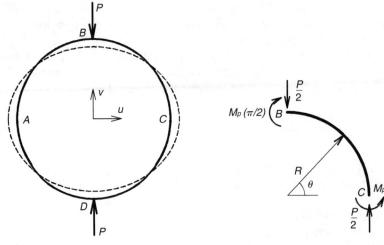

(a) A ring subjected to a pair of vertical loads

(b) Free-body diagram for a quadrant of the ring

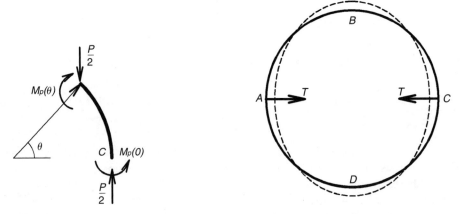

(c) Equilibrium at any section of the ring

(d) The ring subjected to a pair of horizontal forces

Figure 9.1 A ring.

ring, the rotations at points A, B, C and D must be symmetric and only the zero rotations at these locations satisfy the condition. Cutting the ring at any section defined by θ (Figure 9.1c), the bending moment at the section can be written using the equilibrium condition as:

$$M_p(\theta) = M_p(0) + \frac{PR}{2}(1 - \cos\theta) \tag{9.4}$$

where $M_p(\theta)$ at $\theta = 0$ is unknown, but can be determined using the condition that there is no relative rotation between points B and C. Applying a pair of unit moments at B and C to the unloaded quadrant (Figure 9.1b), the bending moment along the ring is a constant:

$$M_{BC}(\theta) = 1 \tag{9.5}$$

Thus the relative rotation between B and C is:

$$\theta_B - \theta_C = \frac{1}{EI}\int_0^{\pi/2} M_P(\theta) M_{BC}(\theta) R d\theta = \int_0^{\pi/2}\left[M_P(0) - \frac{PR}{2}(1-\cos\theta)\right]R d\theta = 0 \tag{9.6}$$

giving:

$$M_P(0) = PR\left(\frac{1}{\pi} - \frac{1}{2}\right) \tag{9.7}$$

Substituting equation 9.7 into equation 9.4 leads to:

$$M_P(\theta) = PR\left(\frac{1}{\pi} - \frac{\cos\theta}{2}\right) \tag{9.8a}$$

Similarly, when the force P is replaced by a unit load, equation 9.8a becomes:

$$\overline{M}_P(\theta) = R\left(\frac{1}{\pi} - \frac{\cos\theta}{2}\right) \tag{9.8b}$$

Substituting equation 9.8 into equation 9.1, the relative vertical deflection between points B and D of the ring is:

$$v_1 = \frac{4}{EI}\int_0^{\pi/2} M_P(\theta)\overline{M}_P(\theta)R d\theta = \frac{PR^3}{EI}\left(\frac{\pi}{4} - \frac{2}{\pi}\right) = 0.1488\frac{PR^3}{EI} \tag{9.9}$$

The deformed shape of the ring subject to the concentrated loads P can thus be formed and is as shown in Figure 9.1a, where points B and D deform inwards while points A and C move outwards. To produce the deformations in the opposite directions to the deformations shown in Figure 9.1a, a pair of horizontal forces, T, are applied at points A and C of the ring in the inward directions as shown in Figure 9.1d. It can be noted that the forcing condition shown in Figure 9.1d can be obtained by rotating the ring and forces shown in Figure 9.1a through 90 degrees anti-clockwise. Thus the bending moments in any section θ of the right top quarter of the ring due to the pair of horizontal forces T and a pair of unit horizontal forces can be written using equation 9.8:

$$M_T(\theta) = TR\left(\frac{1}{\pi} - \frac{\cos(\theta-\pi/2)}{2}\right) = TR\left(\frac{1}{\pi} - \frac{\sin\theta}{2}\right) \tag{9.10a}$$

$$\overline{M}_T(\theta) = R\left(\frac{1}{\pi} - \frac{\sin\theta}{2}\right) \tag{9.10b}$$

The vertical displacement at point B due to the pair of horizontal forces, T, can be evaluated using the second term in equation 9.2 as follows:

$$v_2 = \frac{4}{EI}\int_0^{\pi/2} M_T \overline{M}_p R d\theta = \frac{TR^3}{EI}\left(\frac{1}{2} - \frac{2}{\pi}\right) = -0.1366\frac{TR^3}{EI} \quad (9.11)$$

It can be seen that the pair of forces, T, produce outward displacements between points B and D, which offset some of the displacements induced by the vertical loads P. The negative value is due to the fact that $M_P(\theta)$ ($\overline{M}_P(\theta)$) and $M_T(\theta)$ have opposite signs. Figure 9.2 compares the normalised moments, $M_P(\theta)/PR$ (solid line) and $M_T(\theta)/TR$ (dashed line) along the top right quarter of the ring (between 0 and $\pi/2$).

To provide the pair of forces T, a wire may be added to connect points A and C across the diameter of the ring horizontally as shown Figure 9.3. Again it may be

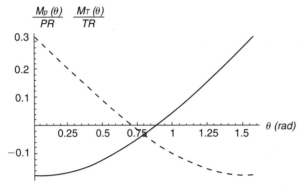

Figure 9.2 Comparison of the normalised bending moments induced by P and T respectively (solid line: $M_P(\theta)/PR$; dashed line: $M_T(\theta)/TR$).

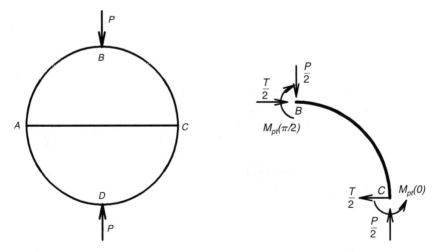

(a) A tied ring subjected to a pair of concentrated vertical loads

(b) Free-body diagram of the upper right part of the tied ring

Figure 9.3 A tied ring.

Smaller internal forces 101

assumed that the tied ring experiences small deformations as was the case for the untied ring. The tied ring shown in Figure 9.3a will be stiffer than the original ring shown in Figure 9.1a. This is because the moments and relative vertical displacement between points B and D of the tied ring become:

$$M_{PT}(\theta) = PR(1/\pi - \cos\theta/2) + TR(1/\pi - \sin\theta/2) \tag{9.12}$$

$$v = v_1 + v_2 = \left(\frac{\pi}{4} - \frac{2}{\pi}\right)\frac{PR^3}{EI} + \left(\frac{1}{2} - \frac{2}{\pi}\right)\frac{TR^3}{EI} = (0.1488P - 0.1366T)\frac{R^3}{EI} \tag{9.13}$$

where the force T can be determined using the compatibility condition for the horizontal displacement between points A and C, i.e. the relative displacement between points A and C of the ring, which is equal to the extension of the wire:

$$u_1 + u_2 + u_T = 0 \tag{9.14}$$

where u_1 and u_2 are the relative horizontal displacements between points A and C induced by the concentrated load P and the horizontal forces T respectively, and u_T is the extension of the wire. The horizontal displacements can be calculated in a similar manner to the vertical displacement:

$$\begin{aligned} u_1 &= \frac{4}{EI}\int_0^{\pi/2} M_P(\theta)\overline{M}_T(\theta)ds \\ &= \frac{4}{EI}\int_0^{\pi/2} PR(\cos\theta/2 - 1/\pi)R(\sin\theta/2 - 1/\pi)Rd\theta \\ &= \frac{PR^3}{EI}\left(\frac{1}{2} - \frac{2}{\pi}\right) \end{aligned} \tag{9.15}$$

$$\begin{aligned} u_2 &= \frac{4}{EI}\int_0^{\pi/2} M_T(\theta)\overline{M}_T(\theta)ds \\ &= \frac{4}{EI}\int_0^{\pi/2} TR^2(\sin\theta/2 - 1/\pi)^2 Rd\theta = \frac{TR^3}{EI}\left(\frac{\pi}{4} - \frac{2}{\pi}\right) \end{aligned} \tag{9.16}$$

The extension of the tie is:

$$u_T = \frac{2TR}{E_T A_T} \tag{9.17}$$

Substituting equations (9.15–9.17) into equation 9.14 leads to:

$$\frac{PR^3}{EI}\left(\frac{1}{2} - \frac{2}{\pi}\right) + \frac{TR^3}{EI}\left(\frac{\pi}{4} - \frac{2}{\pi}\right) + \frac{2TR}{E_T A_T} = 0 \tag{9.18}$$

Introducing the non-dimensional rigidity ratio β:

$$\beta = \frac{E_T A_T R^2}{EI} \qquad (9.19)$$

and substituting equation 9.19 into equation 9.18, the internal force of the wire is:

$$T = \frac{2(4-\pi)\beta}{8\pi + (\pi^2 - 8)\beta} P \qquad (9.20)$$

Equations 9.19 and 9.20 indicate that the tension in the wire is a function of the ratio of the rigidities of the wire to the ring and the radius of the ring. The relationship between T/P and β is plotted in Figure 9.4. It can be seen that T/P increases significantly up to approximately $\beta = 50$ and T/P increases slowly when $\beta > 200$.

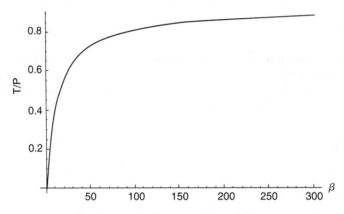

Figure 9.4 The relationship between T/P and β.

Substituting equation 9.20 into equation 9.12 gives:

$$M_{PT}(\theta) = PR(1/\pi - \cos\theta/2) + \frac{2(4-\pi)\beta}{8\pi + (\pi^2-8)\beta} PR(1/\pi - \sin\theta/2)$$

$$= \left[\frac{1}{\pi} - \frac{\cos\theta}{2} + \frac{2(4-\pi)\beta}{8\pi + (\pi^2-8)\beta} \left(\frac{1}{\pi} - \frac{\sin\theta}{2} \right) \right] PR \qquad (9.21)$$

$$= \left[\frac{1}{\pi} - \frac{\cos\theta}{2} + \frac{(2-\pi\sin\theta)(4-\pi)\beta}{(8\pi + (\pi^2-8)\beta)\pi} \right] PR$$

Equation 9.21 indicates that the moment of the tied ring is a function of the location θ and the rigidity ratio β. Figure 9.5 gives the relation between $M_{PT}(\theta)/PR$ and θ when $\beta = 0$, 10 and 100 respectively. It can be observed from Figure 9.5 that the magnitudes of the moments are reduced due to the addition of the tie and with the increase of the rigidity β.

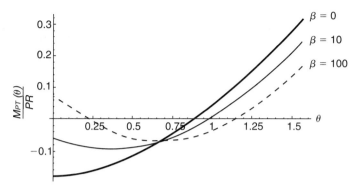

Figure 9.5 Bending moment of the tied ring.

Substituting equation 9.20 into equation 9.13 leads to:

$$v = \left(\frac{\pi}{4} - \frac{2}{\pi}\right)\frac{PR^3}{EI} + \left(\frac{1}{2} - \frac{2}{\pi}\right)\frac{2(4-\pi)\beta}{8\pi + (\pi^2 - 8)\beta}\frac{PR^3}{EI}$$

$$= \left[\left(\frac{\pi}{4} - \frac{2}{\pi}\right) - \frac{(4-\pi)^2 \beta}{8\pi^2 + \pi(\pi^2 - 8)\beta}\right]\frac{PR^3}{EI} \qquad (9.22)$$

The ratio of the relative vertical displacements between points B and D of the tied ring (equation 9.22) to similar displacement of the untied ring (equation 9.9) is:

$$\frac{v}{v_1} = \frac{\left[\left(\frac{\pi}{4} - \frac{2}{\pi}\right) - \frac{(4-\pi)^2 \beta}{8\pi^2 + \pi(\pi^2 - 8)\beta}\right]\frac{PR^3}{EI}}{\left(\frac{\pi}{4} - \frac{2}{\pi}\right)\frac{PR^3}{EI}}$$

$$= 1 - \frac{4\pi(4-\pi)^2 \beta}{(8\pi^2 + \pi(\pi^2 - 8)\beta)(\pi^2 - 8)} \qquad (9.23)$$

Figure 9.6 shows that the displacement ratio changes with the rigidity ratio β. It can

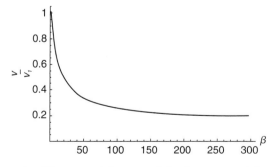

Figure 9.6 The ratio of the vertical displacements of the rings with and without the tie.

be observed from Figure 9.6 that the use of the wire effectively reduces the relative vertical displacement between points B and D.

It has been shown that the larger the tension force in the wire, the smaller the bending moment and the smaller the vertical displacement. On the other hand, the tension force T will not increase significantly when β is larger than 200. Therefore, equation 9.19 can be used to design the rigidity of the tie.

When $E_T A_T \to \infty$ i.e. $\beta \to \infty$, the maximum tension force and the minimum vertical deflection between the points B and D from equation 9.20 and 9.23 respectively become:

$$\lim_{\beta \to \infty} T = \lim_{\beta \to \infty} \frac{2(4-\pi)\beta}{8\pi + (\pi^2 - 8)\beta} P = \frac{2(4-\pi)}{(\pi^2 - 8)} P = 0.9183 P \qquad (9.24)$$

$$\frac{\lim_{\beta \to \infty} v}{v_1} = \frac{\left[\left(\dfrac{\pi}{4} - \dfrac{2}{\pi}\right) - \dfrac{(4-\pi)^2}{\pi(\pi^2 - 8)}\right] \dfrac{PR^3}{EI}}{\left(\dfrac{\pi}{4} - \dfrac{2}{\pi}\right) \dfrac{PR^3}{EI}}$$

$$= 1 - \frac{4(4-\pi)^2}{(\pi^2 - 8)^2} = 1 - 0.8432 = 0.1568$$

(9.25)

Equation 9.25 and Figure 9.6 show that the tied ring is much stiffer than the untied ring and in the extreme case, $\beta \to \infty$, the stiffness of the tied ring is over six times of that of the untied ring.

Example 9.1

A rubber ring has a radius $R = 67.5$ mm and a circular cross-section with diameter $d = 22$ mm as shown in Figure 9.7a. A similar ring has 15 bronze twisting wires tied

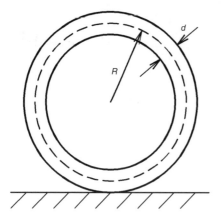

(a) A rubber ring   (b) A tied rubber ring

Figure 9.7 Example 9.1.

horizontally through the centre of the ring, as shown in Figure 9.7b. Each of the wires has a diameter of 0.11 mm. Young's modulus values for the rubber and the bronze wires are 5 N/mm² and 100000 N/mm² respectively. Calculate the vertical displacements of the untied ring and the tied ring when a vertical load P of 22.3 N is applied on the top of the two rings.

Solution

For the rubber ring:

$$EI = \frac{E\pi d^4}{64} = 5 \times \frac{\pi \times 22^4}{64} = 57495 \text{ N mm}^2$$

$$v_1 = 0.1488 \frac{PR^3}{EI} = 0.1488 \frac{22.3 \times 67.5^3}{57495} = 17.7 \text{ mm} \quad (9.9)$$

For the tied rubber ring:

$$E_T A_T = 100000 \times 0.055^2 \pi \times 15 = 14255 \text{ N}$$

$$\beta = \frac{E_T A_T R^2}{EI} = \frac{14255 \times 67.5^2}{57495} = 1130 \quad (9.19)$$

$$v = \left[\left(\frac{\pi}{4} - \frac{2}{\pi} \right) - \frac{(4-\pi)^2 \beta}{8\pi^2 + \pi(\pi^2 - 8)\beta} \right] \frac{PR^3}{EI}$$

$$= \left[\left(\frac{\pi}{4} - \frac{2}{\pi} \right) - \frac{(4-\pi)^2 \times 1130}{8\pi^2 + \pi(\pi^2 - 8) \times 1130} \right] \frac{22.3 \times 67.5^3}{57495} = 2.96 \text{ mm} = 0.167 v_1 \quad (9.22)$$

For the rings, the ratio of the displacements of the tied rubber ring to the untied rubber ring is 0.167. In other words, the vertical stiffness of the tied rubber ring is about six times that of the untied rubber ring. The ratio of 0.167 is slightly larger than 0.157 given in equation 9.25 when the stiffness of the wire is infinite.

9.3 Model demonstrations

9.3.1 A pair of rubber rings

This pair of models demonstrates that *a tied ring is much stiffer than a similar ring without a tie*.

Figure 9.8 shows two rubber rings, one with and one without a wire tied across the diameter. The dimensions and material properties of the rings are described in example 9.1 where the calculated vertical displacements of the rings are given. The same weight of 22.3 N is placed on the top of each of the two rings and the reduced deformation of the tied ring is apparent and its increased stiffness can be seen and felt. This may be explained since the force in the wire increases as the applied load increases, and produces a bending moment (equation 9.10a) in the ring in the

106 *Statics*

opposite direction to the bending moment (equation 9.8a) caused by the external load (Figure 9.2). In this way the force in the wire balances part of the bending moments in the ring, reducing the internal forces in the ring, thus making it stiffer.

Examples of this criterion in practice are tied arches and tied-pitched roofs. The ties help to balance horizontal forces and reduce horizontal displacements, thus effectively increasing the structural stiffness.

Figure 9.8 Comparison of the deformations of two rubber rings.

9.3.2 Post-tensioned plastic beams

This demonstration shows that *a beam with profiled post-tension wires is clearly stiffer than the same beam with straight post-tensioned wires.*

Three 500 mm long plastic tubes are stiffened by a pair of steel wires whose location differs in each tube, see Figure 9.9:

- Model A: the pair of wires is positioned at the neutral plane of the tube.
- Model B: the pair of wires is positioned between the neutral plane and the bottom of the tube.
- Model C: the pair of wires is placed externally with a profiled shape rather than being straight. Two small metal bars, longer than the width of the tube, are placed underneath the tube to create the desired profile for the wires.

Small metal plates are placed at the ends of the tubes to provide the supports to fix the wires. The wires are fixed into screws and the screws pass through the holes in the end plates and are fixed using nuts. By turning the nuts, the tension in the wires and the forces in the tubes are established. Figure 9.9a shows the details of the ends of the three tubes and the locations of the wires. Supporting the beams at their two ends gives three post-tensioned simply supported beams shown in Figure 9.9b. The structural behaviour of the three model beams can be described as follows:

- Model A: as the wires are placed in the neutral plane of the tube, the tube itself is in compression and the wires are in tension.
- Model B: as the wires are placed under the neutral plane of the tube, the tube is

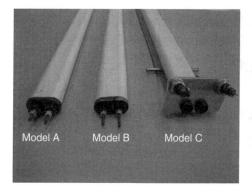 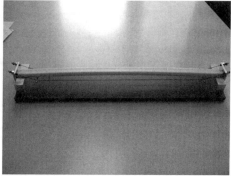

(a) Locations of wires

(b) A plastic beam stiffened by externally profiled wires

Figure 9.9 Three post-tensioned plastic beams with different positions of wires.

subjected to both compression and bending (bending upwards) and the wires are in tension.
- Model C: as shown in Figure 9.9b, the tensions in the wires provide upward forces through the two diverters to the beam, which will partly balance the downward loads. In other words, the upward forces are equivalent to two spring supports to the beam. Thus model C is expected to be significantly stiffer than models A and B.

Positioning the models in turn on the supports shown in Figure 9.9b then pressing at the centre of each beam downwards, it can be felt that model C is obviously stiffer than the models A and B. The three models all form self-balanced systems, but only model C has enlarged bending stiffness. A practical example of a stiffened floor using profiled post-tensioned cables will be given in section 9.4.4.

9.4 Practical examples

9.4.1 Raleigh Arena

The roof structure of the Raleigh Arena (Figure 9.10) consists of carrying (sagging) cables and stabilising (hogging) cables which are supported by a pair of inclined arches. The structure forms, at least in part, a self-balanced system, which effectively reduces the internal forces in the arches. The carrying cables apply large forces to the arches and some of the vertical components of the forces are transmitted to external columns. Significant portions of the bending moments and the horizontal components of the shear and compressive forces in the arches are self-balancing at the points of contact between the two arches. Most of the horizontal components of the remaining shear and compressive forces in the lower parts of the arches are balanced by underground ties, which have a similar function to the wire tie in the ring used in the demonstration. The reduced internal forces not only allow the use of less material but also lead to a stiffer structure.

Another example is the newly built stadium in Nanjing, China. The roof of the stadium is supported by a pair of arches, which are inclined outwards symmetrically

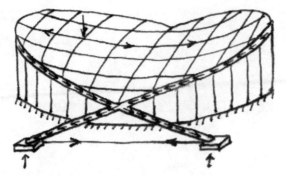

Figure 9.10 The force paths of the Raleigh Arena [9.4].

and do not support each other. Figure 9.11 shows one of the arches. The pair of arches generates large horizontal forces of 13000 kN at their supports. In order to avoid these horizontal forces at the ends of the arches being applied to the pile foundations, which are on soft soil, eight post-tension cables of a diameter of 25 mm and a length of 400 m are placed underground to link the two ends of each arch to balance the large horizontal forces. The cables have the same functions as demonstrated in the tied rubber ring.

Figure 9.11 A stadium in Nanjing, China.

9.4.2 Zhejiang Dragon Sports Centre

The Zhejiang Dragon Sports Centre in China (Figure 9.12) was built in 2000. The stadium has a diameter of 244 m and cantilever roof spans of 50 m, creating unobstructed viewing for the spectators. The roof structure adopts double layer lattice shells that are supported by internal and external ring beams. Cables carry the internal ring beams back to the support towers located at two ends of the stadium. The cables are used as elastic supports to the roof, reducing the internal forces (bending moments) and increasing the stiffness of the shell roof.

The towers are subjected to large forces from the cables, which transmit the

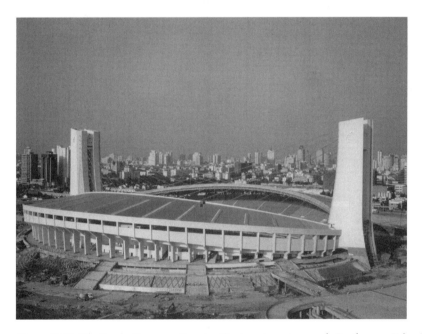

Figure 9.12 Zhejiang Dragon Sports Centre (courtesy of Professor Jida Zhao, China Academy of Building Research, China).

weight of the cantilever roof and the loads on the roof to the towers. These forces in turn cause large bending moments in the 85 m tall cantilever towers. To reduce the bending moments in the towers, post-tension forces were applied to the backs of the towers providing the bending moments opposing the effects of the cables. In this way, the bending moments in the towers become smaller. This leads to a saving of material and a stiffer structure.

110 *Statics*

The bending moments in one of the towers are illustrated in Figure 9.13. Figure 9.13a shows qualitatively the elevation of the tower and the forces applied on the tower together with the bending moments in the tower. The maximum bending moment occurs at the base of the tower and is noted as M_c. Figure 9.13b shows the action of the post-tension forces applied on the tower and the corresponding bending moments. As the distance between the vertical line of action of the compression forces and the neutral axis of the tower varies linearly, the bending moment increases linearly from M_{pa} at the top to M_{pb} at the bottom of the tower, in the opposite direction to those induced by the cable forces. Finally, Figure 9.13c gives the combinations of the two sets of forces on the structure and the resulting bending

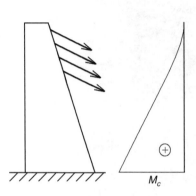

(a) Bending moment due to cable forces

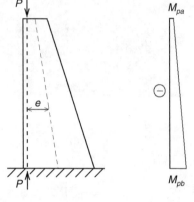

(b) Bending moment due to compression forces

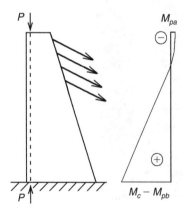

(c) Bending moment due to cable forces and compression forces

Figure 9.13 Illustration of the bending moments in a tower of the Zhejiang Dragon Sports Centre.

moments. Thus the bending moments due to the combined forces are $-M_{pa}$ at the top and $M_c - M_{pb}$ at the base. The reduced bending moments due to the action of the post-tensioning of the tower is obvious. According to equation 9.2, the reduced forces will result in smaller displacements, i.e. a stiffer structure.

9.4.3 A cable-stayed bridge

There are many long-span cable-suspended and cable-stayed bridges in the world. Cables are used for the bridges not only because they are light and have high tensile strength, but also because they can create self-balanced systems in the structures and because the cables provide elastic supports to reduce the effective span of the decks. The latter reason is more significant.

Figure 9.14 shows a cable-stayed bridge in Lisbon. The stayed cables act as elastic supports to the bridge decks, which effectively reduce the internal bending moments in the decks allowing the large clear spans. As the internal forces become smaller, the bridge deck becomes stiffer and can span greater clear distance, producing a more economical design.

This behaviour of the bridge can also be considered using the concept of direct force paths. Due to the use of cables, the loads acting on the bridge decks are not transmitted to their supports primarily through bending actions. Rather, loads are transmitted mainly through the tensile forces in cables to the support tower (Figure 9.14), with the horizontal components of the cable forces induced by the self-weight of the bridge being self-balanced due to the symmetry of the structure. Vertical components of the cable forces pass directly through the tower to the foundation.

As a partially self-balanced system is created, forces are transmitted in a relatively straightforward manner from the deck to the supports, producing a stiffer and more economical design.

Figure 9.14 A cable-stayed bridge in Lisbon.

9.4.4 A floor structure experiencing excessive vibration

A floor in a factory on which machines were operated on a daily basis experienced severe vibrations causing significant discomfort for workers. It was found that resonance occurred when the machines operated. The solution to the problem was to avoid the resonance by increasing the stiffness of the floor or its natural frequency.

It was not feasible to stiffen the floor by positioning additional column supports. However, the resonance problem was solved in a simple and economical manner by the Institute of Building Structures, China Academy of Building Research, Beijing, through stiffening the floor using external post-tensioned tendons as shown in Figure 9.15.

Due to the profile of the tendons and the post-tension forces applied, additional upward forces or elastic supports are provided at the points where the steel bars react against the concrete beams which support the floor. This reduces the internal forces (bending moments) in the beams, making the floor system stiffer with increased natural frequencies. With the natural frequencies avoiding the operating frequency of the machine, resonance did not occur and the response of the floor was significantly reduced. The static behaviour of the structure is demonstrated in section 9.3.2.

Figure 9.15 A floor structure experiencing excessive vibration then stiffened using profiled post-tensioned cables (courtesy of Professor Jida Zhao, China Academy of Building Research, China).

References

9.1 Ji, T. (2003) 'Concepts for designing stiffer structures', *The Structural Engineer*, Vol. 81, No. 21, pp. 36–42.

9.2 Gere, J. M. (2004) *Mechanics of Materials*, Belmont: Thomson Books/Cole.

9.3 Seed, G. M. (2000) *Strength of Materials*, Edinburgh: Saxe-Coburg Publications.

9.4 Bobrowski, J. (1986) 'Design philosophy for long spans in buildings and bridges', *Journal of Structural Engineer*, Vol. 64A, No. 1, pp. 5–12.

10 Buckling

10.1 Definitions and concepts

Buckling of columns: when a slender structural member is loaded with an increasing axial compression force, the member deflects laterally and fails by combined bending and compression rather than by direct compression alone. This phenomenon is called buckling.

Critical load of a structure is the load which creates the borderline between stable and unstable equilibrium of the structure or is the load that causes buckling of the structure.

Lateral torsional buckling of beams: lateral torsional buckling is a phenomenon that occurs in beams which are subjected to vertical loading but suddenly deflect and fail in the lateral and rotational directions.

- The buckling load of a column is proportional to the flexural rigidity of the cross-section EI and the inverse of the column length squared L^2. Increasing the value of the second moment of area of the section I and/or reducing the length L will increase the critical load.

10.2 Theoretical background

10.2.1 Buckling of a column with different boundary conditions

Section 1.3.1 illustrates stable equilibrium and unstable equilibrium using a ruler on two round pens and on one round pen respectively. The stable and the unstable equilibrium can also be illustrated in a different manner. Figure 10.1a shows a rigid bar that has a pin support at its lower end and an elastic spring support with stiffness of k at the other end. An axial compressive load P is applied at the top end of the bar. When the bar is displaced by an amount Δ, there will be a disturbing moment $P\Delta$ and a restoring moment $kL\Delta$ about O. Thus we have:

$P\Delta < kL\Delta \rightarrow$ stable equilibrium

$P\Delta > kL\Delta \rightarrow$ unstable equilibrium

The critical condition occurs when:

$P_{cr}\Delta = kL\Delta$ or $P_{cr} = kL$

P_{cr} is termed the critical load, which is the borderline between stable and unstable equilibrium [10.1].

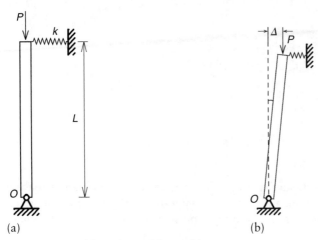

Figure 10.1 Stable and unstable equilibrium.

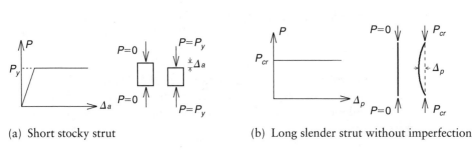

(a) Short stocky strut

(b) Long slender strut without imperfection

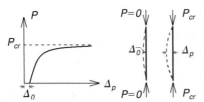

(c) Long slender strut with imperfection

Figure 10.2 Comparison of the behaviour of columns [10.2].

Consider two pin-ended columns that have the same cross-section, but one is short (Figure 10.2a), i.e. its length is not significantly larger than the dimensions of its cross-section, and the other is long (Figure 10.2b), i.e. its length is significantly larger than the dimensions of its cross-section. Normally, they are termed as *stocky* and *slender* struts respectively. When the compressive loads P applied on the ends of the two columns increase, it can be observed that:

- The stocky strut will shorten *axially* by an amount Δ_a, which is proportional to the load applied. When the load reaches P_y, the product of the area of the strut and the compressive yield stress of the material, the material will deform plastically. The column 'squashes' and P_y is termed as the *squash load*. After removing the load, the column retains some permanent deformation. The relationship between the load and the axial deformation is characterised in Figure 10.2a.
- The slender strut without imperfection will remain straight for a perfect strut when P is small. When the load reaches the critical load P_{cr}, the product of the area of the strut and the critical stress, the strut will suddenly move excessively in a lateral direction. This mode of failure is termed *buckling* in which the critical stress is elastic and is often smaller or much smaller than the elastic limit stress. The relationship between the load and the lateral deformation at the mid-height of the strut is characterised in Figure 10.2b.
- In practice a slender strut is unlikely to be perfectly straight. If an initial imperfection is considered to be defined by a lateral deformation at the mid-height of the strut with a value of Δ_0 under zero axial load, the lateral deformation will grow at an increasing rate as the load increases. At a certain load which is normally smaller than or equal to the critical load P_{cr}, the member fails due to excessive lateral deformation. The relationship between the load and deformation is characterised in Figure 10.2c.

Clearly the short stocky strut and the long slender strut have different modes of failure. The former depends on the yield stress of the material in compression and the area of the section but is independent of the elastic modulus, the length of the strut and the shape of the section, while the latter is independent of the yield stress but depends on the length of the strut, the elastic modulus and the shape and area of the section.

Figure 10.3a shows a column with pin joints at each end and it is assumed that the column is straight when unloaded. An axial compressive load is applied through the longitudinal axis of the column with increasing magnitude until the column takes up the deformed shape as shown in Figure 10.3b. For this deformation state

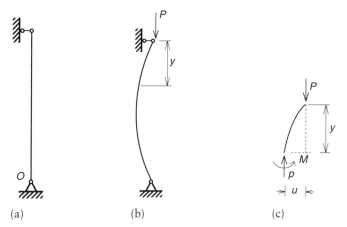

(a)   (b)   (c)

Figure 10.3 Buckling of a pin-ended column.

the equilibrium equation is to be established to capture the nature of buckling phenomenon. If Figure 10.3b rotates anti-clockwise through 90 degrees, it becomes a beam in bending subject to a pair of end forces. Thus the theory of beam bending in Chapter 7 can be used to establish basic equations for the behaviour of the member. In order to obtain the differential equation of equilibrium, consider a free-body diagram from the beam as shown in Figure 10.3c. At the distance y from the top joint, the displacement is u, the bending moment is $M = Pu$ and the compressive force is P. These forces and the load P at the top maintain the free-body in equilibrium. Using equation 7.2 gives:

$$EI\frac{d^2u}{dx^2} = -Pu \quad \text{or} \quad \frac{d^2u}{dx^2} + \frac{P}{EI}u = 0 \tag{10.1}$$

Solving equation 10.1 [10.1, 10.3] gives the expression of the buckling load of the simply supported column as follows:

$$P_{cr} = \frac{\pi^2 EI}{L^2} \tag{10.2}$$

Equation 10.2 indicates that *the buckling load is proportional to the rigidity of the cross-section EI and the inverse of the column length squared L^2*. Increasing the value of I and/or reducing the length L will increase the critical load.

The buckling load of the column with different boundary conditions can be derived in a similar manner. The expressions for the buckling loads will have the same form as equation 10.2 and can be shown in a unified formula:

$$P_{cr} = \frac{\pi^2 EI}{(\mu L)^2} \tag{10.3}$$

where μL is the effective length of the column considering different boundary conditions. Table 10.1 lists and compares the buckling loads of a column with four different boundary conditions.

The results in Table 10.1 show that the critical load of a slender strut with two fixed ends is four times that of a simply supported column and the critical load of a simply supported column is four times that of a cantilever. When the compressive load on a slender strut approaches its critical load, the effective stiffness of the column tends to zero.

More detailed information for buckling of a column can be found in [10.1, 10.2 and 10.3].

10.2.2 Lateral torsional buckling of beams

The primary function of a beam carrying vertical loading is to transfer loads by means of bending action and the beam deforms within its vertical plane until it reaches the full plastic yield stress over the whole of its section at failure. This is only true if the beam is restrained from sideways or out-of-plane movement. If no lateral restraint is provided, the beam under load bends in the vertical plane passing through its neutral axis. Gradually increasing load results in a situation where the

Buckling 117

Table 10.1 Comparison of buckling loads of a column with different boundary conditions

Boundary conditions	Pinned–pinned	Fixed–fixed	Fixed–pinned	Fixed–free
Critical load	$\dfrac{\pi^2 EI}{L^2}$	$\dfrac{4\pi^2 EI}{L^2}$	$\dfrac{2.05\pi^2 EI}{L^2}$	$\dfrac{\pi^2 EI}{4L^2}$
Effective length μL	L	$0.5L$	$0.7L$	$2L$
Relative critical load	1	4	2.05	0.25

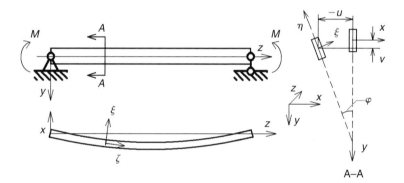

Figure 10.4 Lateral buckling of a rectangular beam.

beam may suddenly deflect sideways and twists before the full yield stress is reached. This mode of failure is known as *lateral torsional buckling*, in which collapse is initiated as a result of lateral deflection and twisting [10.2, 10.4 and 10.5].

Consider a uniform, straight and elastic beam without any initial bow or twist (Figure 10.4), simply supported in both the y–z plane and x–z plane. A pair of moments is applied at the two ends of the beam and in the y–z plane of the beam with increasing magnitude until the beam takes up the deformed shape shown in Figure 10.4. For this deformation state the equilibrium equation is to be established to capture the nature of lateral torsional buckling phenomenon. Figure 10.4 also shows the geometrical relationship of section A–A before and after lateral deformation, illustrating that lateral buckling involves both a lateral deflection u and a twist

φ about an axis parallel to the z axis. Both deformations need to be considered to determine the critical load.

The differential equation of equilibrium is [10.2]:

$$\frac{d^2\varphi}{dx^2} + \frac{M^2}{EI_y GJ}\varphi = 0 \tag{10.4}$$

where I_y is the second moment of area of the beam in respect to the y axis. G and J are the shear modulus and polar second moment of area or torsional constant of the section, which have been defined in Chapter 5. For a rectangular section $J = hb^3/3$ in which b and h are the width and height of the beam.

Equation 10.4 has the same format as equation 10.1. Using the same solution process or simply the analogy, the critical moment, similar to the critical load for buckling of a column, can be expressed as:

$$M_{cr} = \frac{\pi\sqrt{EI_y GJ}}{L} \tag{10.5}$$

Equation 10.5 shows the coupled nature of the deformation involved in the buckled shape (Figure 10.4) due to the presence of EI_y, the lateral bending rigidity, and GJ, the torsional rigidity.

For rectangular beams with depth h and width b, $I_y = hb^3/12$ and $J = hb^3/3$ and substituting them into equation 10.5 gives:

$$M_{cr} = \frac{\pi h b^3 \sqrt{EG}}{6L} \tag{10.6}$$

Equation 10.6 indicates that *the buckling moment is proportional to the height and the third power of the width of the beam and the inverse of the span of the beam.* Therefore increasing the width of the beam will be more effective than increasing the height for increasing the critical moment.

Equation 10.5 or equation 10.6 is obtained when the beam is subjected to a constant bending moment. When a concentrated load is applied at the centroid of the free end of a cantilever, the bending moment in the vertical plane at any cross-section with distance z from the fixed origin is:

$$M = P(L - z) \tag{10.7}$$

Substituting equation 10.7 into equation 10.4 and solving the equation leads to [10.4]:

$$M_{cr} = 1.28\frac{\pi\sqrt{EI_y GJ}}{L} = 0.67\frac{hb^3\sqrt{EG}}{L} \tag{10.8}$$

Equation 10.8 will be verified in section 10.3.3.

Example 10.1

Consider three brass cantilever beams, A, B and C, which have the same length of 270 mm but have different section sizes, 12.7 mm×0.397 mm (1/2 in.×1/64 in.), 12.7 mm×0.794 mm (1/2 in.×1/32 in.) and 25.4 mm×0.397 mm (1 in.×1/64 in.). The elastic modulus and shear modulus of the material are 100 *GPa* and 40 *GPa* respectively [10.3]. Calculate the buckling moments of the three beams and the maximum concentrated loads which may be applied at the free ends of the beams.

Solution

Equation 10.8 can be directly used for calculating the critical moment of cantilever beam A as follows:

$$M_{cr} = 0.67 \frac{hb^3 \sqrt{EJ}}{L} = \frac{0.67(12.7)(0.397)^3 \sqrt{100000 \times 40000}}{270} = 125 \text{ Nmm}$$

The critical load at the free end of the cantilever beam is:

$$P_{cr} = M_{cr}/L = 125/270 = 0.462 \text{ N}$$

Beam B has twice the thickness of beam A and beam C has twice the height of beam A; the critical loads for beams B and C can be determined from equation 10.8 and are 3.70 N and 0.924 N respectively. The three beams used in this example will be tested in section 10.3.3 to examine the predictions from equation 10.8.

When an I-sectioned beam is subjected to torsion, axial deformation of the flange will occur. This type of deformation is called *warping*. Therefore, the applied torque will be resisted by the shear stresses associated with pure torsion and the axial stresses associated with warping. In other words, considering the effect of warping, the member will have a larger capacity to resist the external torque or will have a larger buckling moment.

When the beam with the narrow rectangular section studied above is replaced by an I-section, the effect of warping needs to be considered and equation 10.5 is extended to [10.2]:

$$M_{cr} = \frac{\pi \sqrt{EI_y GJ}}{L} \sqrt{1 + \frac{\pi^2 EI_w}{L^2 GJ}} \quad (10.9)$$

where $I_w = I_y h^2/4$ is the warping constant. If $I_w = 0$, equation 10.9 reduces to equation 10.5 and considering the effect of warping increases the buckling moment.

10.3 Model demonstrations

10.3.1 Buckling shapes of plastic columns

These models demonstrate *the phenomenon of buckling and the buckling shapes of a column with different boundary conditions. Users can feel the magnitudes of the buckling forces of the plastic columns.*

120 *Statics*

Column buckling behaviour can be demonstrated using slender members such as a thin plastic ruler.

1. Put one end of the ruler on the surface of a table and the other end in the palm of one's hand. Press axially on the top end of the ruler and gradually increase the compression force. The straight ruler will suddenly deflect laterally as shown in Figure 10.5a. The deformation becomes larger with further application of the compressive force. This simulates the buckling of a column with two pinned ends.
2. Now hold the two ends of the ruler tightly to prevent any rotational and lateral movements of the two ends of the ruler. Then gradually press axially on the ruler using the fingers until the ruler deforms sideways as shown in Figure 10.5b. This demonstrates a different buckling shape for a column with two ends fixed. One can clearly feel that a larger force is needed in this demonstration than in the previous demonstration.
3. If one intermediate lateral support is provided, so that the ruler cannot move sideways at this point, a larger compressive force will be required to make the ruler buckle in the shape shown in Figure 10.5c.

(a) Two pinned ends (b) Two fixed ends (c) Two pinned ends with a lateral support at the middle of the rule

Figure 10.5 Buckling shapes and boundary conditions of a plastic ruler.

10.3.2 Buckling loads and boundary conditions

This model demonstrates *qualitatively and quantitatively the relationships between the critical loads and boundary conditions of a column (equation 10.3) and the corresponding buckling shapes*.

The model shown in Figure 10.6 contains four columns with the same cross-section but each with different boundary conditions; (from left to right and from bottom to top), pinned–pinned, fixed–fixed, pinned–fixed and free–fixed. The model has load platens on the tops of the columns which allow weights to be added and

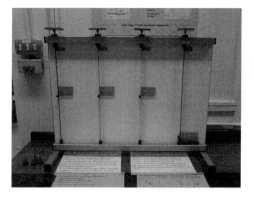 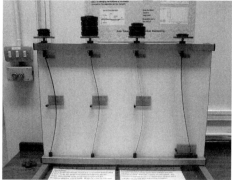

(a) Four columns with different boundary conditions

(b) Critical loads and buckling shapes

Figure 10.6 Critical loads, buckling shapes and boundary conditions of columns.

applied to the columns. This model can demonstrate the different buckling or critical loads of the columns and the associated buckling shapes or modes.

1. The different boundary conditions should be observed by the user before applying loads to the columns (Figure 10.6a).
2. Weights are added incrementally to the platens until the columns buckle (Figure 10.6b) at which time the loads, including the weights and the platens, are the critical loads of the columns and the shapes of the columns are the buckling modes.
3. With all columns loaded (Figure 10.6b), the relative buckling loads and buckling modes can be observed and compared for the different boundary conditions. Table 10.2 gives the measured critical loads of the four columns.

It is to be expected that there will be some differences between the predicted and the test critical loads, but significant error occurs in the fixed–free model (Table 10.2). Several trials showed that the free-end at the bottom of the strut was not really free to move laterally due to a small amount of friction. In addition, the column is shorter than the others, which can be seen in Figure 10.6a. Real structures rarely behave in exactly the same manner as the theoretical models.

Table 10.2 Comparison of buckling loads of a column with different boundary conditions

Boundary conditions	Pinned–pinned	Fixed–fixed	Fixed–pinned	Fixed–free
Critical load	$\dfrac{\pi^2 EI}{L^2}$	$\dfrac{4\pi^2 EI}{L^2}$	$\dfrac{2.05\pi^2 EI}{L^2}$	$\dfrac{\pi^2 EI}{4L^2}$
Relative critical load	1	4	2.05	0.25
Test critical load (N)	5.61	20.8	13.0	4.05
Relative test critical load	1	3.71	2.31	0.72

10.3.3 Lateral buckling of beams

This set of models demonstrates *the behaviour of lateral buckling of a narrow rectangular beam with different sizes of section and thus the validity of equation 10.8*.

The buckling behaviour of the three brass cantilevers described in example 10.1 is demonstrated through simple tests. One end of each of the brass strips is fixed to a wooden block through screws, creating cantilevers as shown in Figure 10.7.

Model A: 270 mm long cantilever with a 12.7 mm × 0.397 mm rectangular cross-section

Hold one end of the cantilever (Figure 10.7a) firmly and apply a vertical concentrated load at the free end. Increase the load gradually until the beam moves sideways and twist (when the loading is about 0.49 N). This type of deformation typifies the form of instability called lateral torsional buckling.

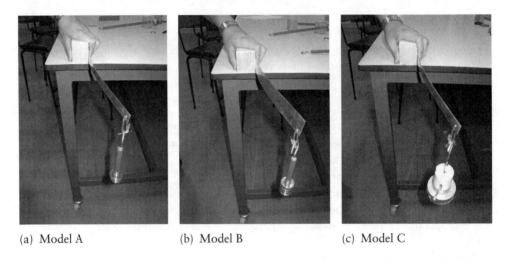

(a) Model A (b) Model B (c) Model C

Figure 10.7 Lateral torsional buckling behaviour.

Model B: 270 mm long cantilever with a 25.4 mm × 0.397 mm rectangular cross-section

The effect of increasing the height of the section: repeat the type of test carried out for model A. When the concentrated load reaches approximately 0.98 N, the cantilever starts to move sideways and twist (Figure 10.7b). This value is twice that of the critical load for model A. As predicted by equation 10.8, doubling the height of the narrow beam doubles the buckling moment or the critical load.

Model C: 270 mm long cantilever with a 12.7 mm × 0.794 mm rectangular cross-section

The effect of increasing the width of the section: repeat the type of test carried out for models A and B. When the concentrated load reaches approximately 3.82 N, the cantilever starts to move sideways and twist (Figure 10.7c). The load is now nearly eight times that which caused lateral torsional buckling of model A. It is about four times that which caused lateral torsional buckling of model B although the cantilever uses the same amount of material as model B. Following equation 10.8, the buckling load of model C should be exactly eight times that of model A and four times that of model B.

Model D: model A with a lateral support

The effect of lateral restraint: repeat the type of test carried out for model A but this time hold the cantilever at mid-span using a finger and thumb to prevent lateral movement. The loading can increase significantly without lateral buckling. This indicates that the lateral restraint effectively increases the lateral torsional buckling capacity of the cantilever.

Table 10.3 summarises the theoretical values obtained from equation 10.8 and the experimental values of the lateral buckling loads. There is close agreement between the two sets of values.

Table 10.3 Comparison between the calculated and measured lateral buckling loads

Boundary conditions	Model A	Model B	Model C
Theoretical critical load (N)	0.462	0.924	3.7
Relative theoretical critical load	1	2	8
Test critical load (N)	0.49	0.98	3.83
Relative test critical load	1	2	7.8

10.3.4 Buckling of an empty aluminium can

This demonstration shows that local *buckling occurred on a drinks can causes the global collapse of the can*.

The phenomenon of buckling is not limited to columns. Buckling can occur in many kinds of structures and can take many forms. Figure 10.8a shows that two

(a)   (b)

Figure 10.8 Buckling failure of empty drinks cans.

124 *Statics*

empty drinks cans can carry a standing person of 65 kg. However, this is close to the critical loads of the two cans. When the standing person slightly moved his body, which caused a small shift of some of the body weight from one can to the other, the thin cylindrical wall of one can buckled and the cans then collapsed completely as shown in Figure 10.8b.

10.4 Practical examples

10.4.1 Buckling of bracing members

Steel bracing members are normally slender and are not ideal for use in compression. Therefore they are usually designed to transmit tension forces only. For cross-braced panels which are often used to resist lateral loads, one bracing member will be in tension and the other in compression. When the loading is applied in the opposite direction, the member previously in tension will be subjected to compression and the other member becomes a tension member.

Figure 10.9a shows one of two bracing members in a panel of a storage rack which has buckled. Comparing the value of the material stored and cost of the bracing members, the use of more substantial bracing members would have easily been justified. Figure 10.9b shows two buckled bracing members in a building.

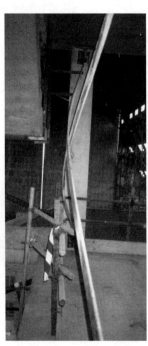

(a) Buckling of a bracing member in a storage rack (courtesy of Dr A. Mann, Jacobs)

(b) Buckling of two bracing members in a building

Figure 10.9 Buckling of bracing members.

10.4.2 Buckling of a box girder

Figure 10.10 Buckling of a box girder (courtesy of Dr A. Mann, Jacobs).

Figure 10.10 shows a box girder that buckled during an earthquake. The girder was subjected to large compressive forces and the material was also squashed.

10.4.3 Prevention of lateral buckling of beams

Figure 10.11 Additional supports provided to prevent lateral torsional buckling (courtesy of Westok Ltd).

Figure 10.11 shows a system where additional members are provided to prevent the lateral and torsional buckling of cantilever beams through reducing the beam length. Cellular beams are used as supporting structures for the cantilever roof of a grandstand. The lower parts of the beams are subjected to compressive forces, but the lateral supports from the roof cladding on the top of the beams may not be effective at preventing lateral and torsional buckling at the bottom parts of the beams, thus additional members are placed perpendicular to the beams.

Figure 10.12 shows the lateral supports provided to two arch bridges. The loads on the bridges are transmitted through mainly the compression forces in arches to the supports of the arches. Thus lateral supports are provided to prevent any possible lateral buckling.

Figure 10.12 Members are provided to prevent the lateral buckling of arches.

References

10.1 Benham, P. P., Crawford, R. J. and Armstrong, C. G. (1998) *Mechanics of Engineering Materials*, Harlow: Addison Wesley Longman Ltd.
10.2 Kirby, P. A. and Nethercot, D. A. (1979) *Design for Structural Stability*, London: Granada Publishing.
10.3 Gere, J. M. (2004) *Mechanics of Materials*, Sixth Edition, Belmont: Thomson Books/Cole.
10.4 Allen, H. G. and Bulson, P. S. (1980) *Background of Buckling*, New York: McGraw-Hill.
10.5 Timoshenko, S. P. and Gere, J. M. (1961) *Theory of Elastic Stability*, New York: McGraw-Hill.

11 Prestress

11.1 Definitions and concepts

Prestressing is a technique that generates stresses in structural elements before they are loaded. This can be used to reduce particular unwanted stresses and/or displacements which would develop due to external loads, or for generating particular shapes of tension structures.

Prestressing techniques can be used to achieve:

- A redistribution of internal forces: reduction of the maximum internal forces will permit the design of lighter structures.
- Avoidance of cracks: keeping a member in constant compression means that there will be no cracks.
- Stiffening a structure or a structural element.

11.2 Theoretical background

Barrels were, and still are, made from separated wooden staves, which are kept in place by metal bands/hoops as shown in Figure 11.1a. The metal bands are slightly smaller in diameter than the diameter of the barrel, and are forced into place over

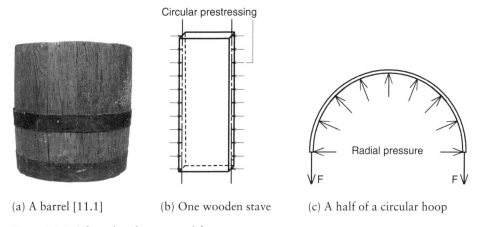

(a) A barrel [11.1]   (b) One wooden stave   (c) A half of a circular hoop

Figure 11.1 A barrel and its internal forces.

the staves. This forces the staves together forming a watertight barrel, i.e. a structure. From free-body diagrams of a typical stave and half the metal band, as shown in Figures 11.1b and 11.1c, it can be seen that circular prestressing (compression) forces are applied along the vertical sides of the stave and the tensile force F on half of the metal band due to internal pressure, and this is balanced by circular hoop pressure. When liquid is added to the barrel, the liquid applies tension forces along the sides of each stave, but this force is and must be smaller than the pre-added compressive forces otherwise the barrel will leak.

Pretensioning and/or post-tensioning is widely used in concrete elements, creating prestressed concrete. Prestressed concrete may be considered as essentially a concrete structure with the tendons supplying the prestress to the concrete, or as steel and concrete acting together, with steel taking tension and concrete taking compression so that the two materials form a resisting couple against the external actions [11.2, 11.3]. Here focus is on the first aspect.

Concentrically prestressed beams

Consider a simply supported beam prestressed by a tendon through its neutral axis and loaded by external loads, as shown in Figure 11.2a. Due to the pretension force F, a uniform compressive stress σ_c occurs across the section of an area A and this is:

$$\sigma_c = \frac{F}{A} \tag{11.1}$$

The stress distribution is shown in Figure 11.2c. If M is the maximum moment at the centre of the beam induced by the external load, the normal stress at any point y across the section is:

$$\sigma_b = \frac{My}{I} \tag{11.2}$$

where y is the distance from the neutral axis and I is the second moment of area of the section about its neutral axis. The stress distribution defined in equation 11.2 is illustrated in Figure 11.2d. Thus the resulting normal stress distribution on the section is:

$$\sigma = \frac{F}{A} \pm \frac{My}{I} \tag{11.3}$$

which is shown in Figure 11.2e. If there is no tensile stress in any section for the given prestress and load conditions, a beam comprising separate blocks and a tendon shown in Figure 11.2b is similar to the beam in Figure 11.2a. The prestress provides compressive stress on the sections of the beam which removes or reduces the tensile stress induced by external loads.

Eccentrically prestressed beams

If the location of the tendon in Figure 11.2a is placed eccentrically with respect to the neutral axis of the beam by a distance of e as shown in Figure 11.3a, an addi-

Prestress   129

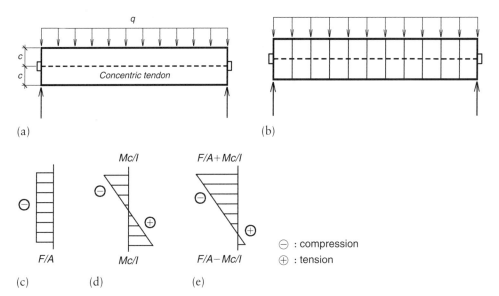

Figure 11.2 A centrally prestressed beam.

tional moment will be induced due to the eccentricity e and the resulting normal stress is given as follows:

$$\sigma_e = \frac{Fey}{I} \qquad (11.4)$$

The normal stress distribution across a typical section is illustrated in Figure 11.3d. It can be seen from Figures 11.3c and 11.3d that σ_e is in the opposite direction to

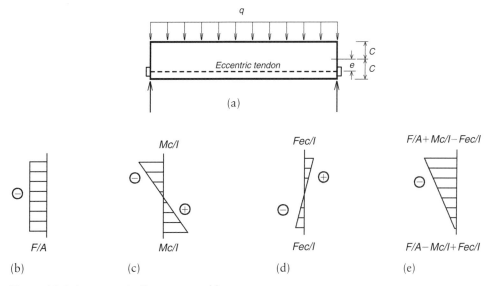

Figure 11.3 An eccentrically prestressed beam.

130 Statics

that induced by the load (equation 11.2), making the stress distribution across the section more even.

Externally prestressed beams

Prestressing tendons can also be used externally and the tendons can be bent or curved. Figure 11.4a shows a similar beam with two tendons with two bends placed externally and symmetrically. If a free-body diagram is drawn for the beam and the tendons, it can be seen that the tendons provide not only the direct compressive forces but also a pair of upward forces (Figure 11.4b), which can effectively balance part of external load. To simplify the discussion, it is assumed that there is no friction loss along the tendon due to the sharp bend and that the deviation produced by the bends is small in comparison with the length of the beam [11.2, 11.3]. The upward force P is equal to the component of the pretension force in the vertical direction, F_y. The resulting normal stress σ_p due to the pair of vertical upward forces on the beam is:

$$\sigma_p = \frac{F_y a y}{I} \qquad (11.5)$$

Example 11.1

A simply supported prestressed concrete rectangular beam with a span of 8 m, a width of $b = 0.3$ m and a height of $2c = 0.6$ m, is subjected to uniform vertical loading of $q = 20$ kN/m. Three ways of using prestressing are considered as follows:

- Case A: a prestressing tendon is placed at the neutral axis of the beam with a force of 200 kN as shown in Figure 11.2a.
- Case B: a prestressing tendon is placed at $e = 0.2$ m below the neutral axis with a force of 200 kN as shown in Figure 11.3a.
- Case C: a pair of prestressing tendons are placed with bends at $a = 1.5$ m and $c = 0.3$ m as shown in Figure 11.4a. The total forces of the two tendons are 200 kN.

Calculate the maximum and minimum stresses in the central section of the beam for the three designs.

Solution

The horizontal and vertical components of the tendon force applied at the ends of the beam for case C are respectively (Figure 11.4b):

$$F_x = 200 \times \frac{1.5}{\sqrt{1.5^2 + 0.3^2}} = 196 \text{ kN}$$

$$F_y = P = 200 \times \frac{0.3}{\sqrt{1.5^2 + 0.3^2}} = 39.2 \text{ kN}$$

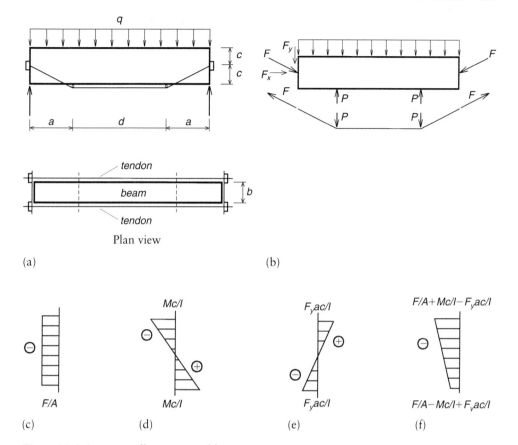

Figure 11.4 An externally prestressed beam.

The moments induced by the uniform load, the eccentricity and the upward forces are thus:

$$M = \frac{qL^2}{8} = \frac{20 \times 8^2}{8} = 160 \, \text{kNm}$$

$$M_e = Fe = 200 \times 0.2 = 40 \, \text{kNm}$$

$$M_p = F_y a = 39.2 \times 1.5 = 58.8 \, \text{kNm}$$

Thus the maximum normal stresses induced by F (or F_x), M, M_e and M_p at the centre of the beam are respectively:

$$\sigma_c = \frac{F}{A} = \frac{200000}{300 \times 600} = 1.11 \, \text{N/mm}^2$$

$$\sigma_{ct} = \frac{F_x}{A} = \frac{196000}{300 \times 600} = 1.09 \, \text{N/mm}^2$$

$$\sigma_q = \frac{Mc}{bh^3/12} = \frac{160 \times 0.3 \times 10^9 \times 12}{0.3 \times 0.6^3 \times 10^{12}} = 8.89 \, \text{N/mm}^2$$

$$\sigma_e = \frac{M_e c}{bh^3/12} = \frac{40 \times 0.3 \times 10^9 \times 12}{0.3 \times 0.6^3 \times 10^{12}} = 2.22 \, \text{N/mm}^2$$

$$\sigma_p = \frac{M_p c}{bh^3/12} = \frac{58.8 \times 0.3 \times 10^9 \times 12}{0.3 \times 0.6^3 \times 10^{12}} = 3.27 \, \text{N/mm}^2$$

Thus the resulting stresses at the bottom and top fibres in the section at the centre of the beams are:

Case A:

Bottom: $-\sigma_{c1} + \sigma_q = -1.11 + 8.89 = 7.78 \, \text{N/mm}^2$

Top: $-\sigma_{c1} - \sigma_q = -1.11 - 8.89 = -10.0 \, \text{N/mm}^2$

Case B:

Bottom: $-\sigma_c + \sigma_q - \sigma_e = -1.11 + 8.89 - 2.22 = 5.56 \, \text{N/mm}^2$

Top: $-\sigma_c - \sigma_q + \sigma_e = -1.11 - 8.89 + 2.22 = -7.78 \, \text{N/mm}^2$

Case C:

Bottom: $-\sigma_{ct} + \sigma_q - \sigma_p = -1.09 + 8.89 - 3.27 = 4.53 \, \text{N/mm}^2$

Top: $-\sigma_{ct} - \sigma_q + \sigma_p = -1.09 - 8.89 + 3.27 = -6.71 \, \text{N/mm}^2$

The results show that the eccentrically placed tendon is more effective than the centrally placed tendon in reducing the stress levels; and the externally placed profiled tendons may be even more effective. The other benefit of using the externally placed profiled tendons is to increase the stiffness of the beam. A practical example is given in section 9.4.4.

Prestressing can also be used to create tension structures in which the internal forces of the structures are tensile; for example fabric membranes, prestressed cable nets and cable beams in the form of trusses or girders.

High-tensile steel cables can transmit large axial forces in tension. The ever-increasing spans and elegance of modern suspension and cable-stayed bridges and cable roofing structures are the most obvious examples in which large loads are supported and transmitted by members and cables in tension. The Raleigh Arena in North Carolina, US, mentioned in section 9.4.1, is well-known for its cable roof structure supported by a pair of inward-inclined arches [11.4, 11.5].

Prestress 133

11.3 Model demonstrations

11.3.1 Prestressed wooden blocks forming a beam and a column

This model demonstrates *the effect of prestressing which makes separate wooden blocks act as a beam or column.*

(a) (b)

Figure 11.5 Effect of prestressing (1).

Figure 11.5a shows a number of separated wooden cubes which are linked using a metal wire through small holes at their centres. One end of the wire passes over a support post while the other end is constrained horizontally by a metal post preventing the cubes from falling down. The structure formed cannot support even its own weight and the cubes simply hang on the wire. If weights are attached to the right-hand end of the wire, the wire is tightened. The metal post and the increased compressive prestress between the cubes enable them to form a structure that can now carry an external load as shown in Figure 11.5b.

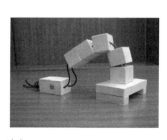 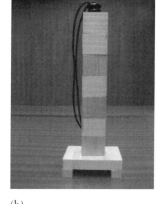 

(a) (b) (c)

Figure 11.6 Effect of prestressing (2).

134 *Statics*

A loose elastic string, with one end fixed to a base, passes through the central holes of a pile of wooden blocks as shown in Figure 11.6. The effect of prestress can be demonstrated as follows:

1 Push the column from one side, it will topple as shown in Figure 11.6a.
2 Reform the column and tighten the elastic string anchoring it to the top block (Figure 11.6b)
3 Hold the base of the model and again push one side of the column. This time the blocks act as a single member which cannot be easily toppled as shown in Figure 11.6c.

11.3.2 A toy using prestressing

This model demonstrates *the effect of tension in strings which makes a toy stable and upright.*

A popular toy (Figure 11.7a) makes use of prestressing. The mechanism for the prestressing is illustrated in Figures 11.7b and 11.7c.

Strings are threaded through the legs of the toy and attached to the base. A spring is used to push the base down and create tension in the strings to hold the toy upright and stable (Figure 11.7b).

When the base is pushed upwards, the tension in the string is released. The sections of the legs are no longer held in place and become unstable and the toy collapses (Figure 11.7c). When the force is released, the spring makes the base move down, the strings become taught and the toy straightens and becomes stable.

11.4 Practical examples

11.4.1 A centrally post-tensioned column

Figure 11.8 shows three piles of stones which form stone columns in a park. This landmark is similar to the wooden block column model shown in Figure 11.6. A

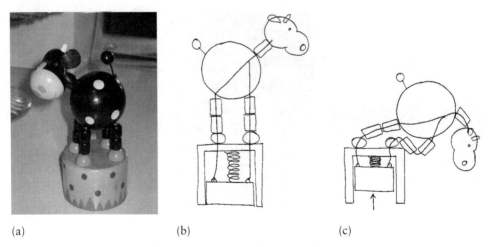

(a)　　　　　　　　　(b)　　　　　　　　　(c)

Figure 11.7 Prestressing used in a toy (courtesy of Miss G. Christian) [11.6].

steel bar, with one end anchored into a foundation below the stones, is threaded through pre-made central holes in the stones. The bar is then tensioned to make the stone blocks act together similar to a single member.

Figure 11.8 Centrally post-tensioned stone columns.

11.4.2 An eccentrically post-tensioned beam

In order to reduce the maximum sagging bending moment of a beam and its maximum deflection when loaded, prestressing can be used to produce initial hogging bending moments and upward deflections, which will offset parts of the bending moments and deflections induced by the subsequent downward vertical loading.

To produce a hogging bending moment and upward deflection of a beam, the post-tensioned steel bar needs to be placed below the neutral axis of the cross-section of the beam. The position of the tendon should be as low as possible to produce larger hogging moments and deflections or to allow smaller post-tension forces.

Figure 11.9 shows a composite beam that has a span of over 10 m [11.7]. The cross-section of the beam is T-shaped and the post-tensioning tendon is placed at a position much lower than the neutral axis of the section.

11.4.3 Spider's web

Spiders have been able to produce silk for the last 400 million years and have been building orb webs for the last 180 million years. The webs have evolved to arrest the

136 *Statics*

Figure 11.9 An eccentrically post-tensioned beam.

flight and capture fast-moving and relatively large insects [11.8, 11.9]. Figure 11.10 shows two spiders' webs that consist of radial threads, capture spirals and anchor threads.

The anchor threads span from the supports and suspend the radial threads. The radial threads are then overlaid by the sticky capture spirals. The prestressing is

Figure 11.10 Spiders' webs (courtesy of Dr A. S. K. Kwan, Cardiff University).

applied during construction by simply pulling the individual strands tighter around their supports.

11.4.4 A cable-net roof

Cable-net roofs are ideally suited for use with long span structures such as gymnasia. Figure 11.11a shows half of the cable-net roof used for the Sichuan Provincial Gymnasium, which was built in 1988 and is able to contain over 10000 seated spectators [11.10].

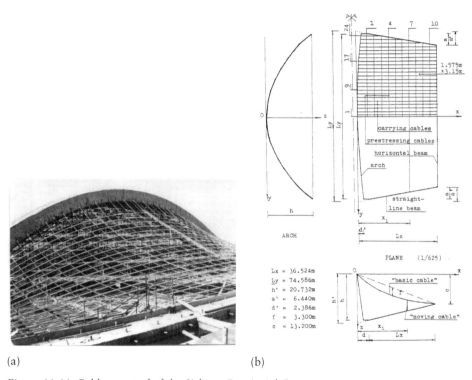

(a)   (b)

Figure 11.11 Cable-net roof of the Sichuan Provincial Gymnasium.

The stiffness of the cable-net roof was established by prestressing. By applying forces at the ends of prestressing (hogging) cables, tension forces are induced in both carrying (sagging) cables and prestressing cables. The prestressing and carrying cables are normally placed perpendicularly to each other (Figure 11.11b). The level of prestress applied should ensure that no prestressing cables are slack for any load configuration.

References

11.1   Hemera Technologies (2001) *Photo-Objects 50,000 Volume 2*, Canada.
11.2   Lin, T. Y. (1955) *Design of Prestressed Concrete Structures*, New York: John Wiley & Sons.

11.3 Nawy, E. G. (1995) *Prestressed Concrete, A Fundamental Approach*, Second Edition, New Jersey: Prentice-Hall.
11.4 Buchholdt, H. A. (1985) *Introduction to Cable Roof Structures*, Cambridge: Cambridge University Press.
11.5 Lewis, W. J. (2003) *Tension Structures: Form and Behaviour*, London: Thomas Telford.
11.6 Ji, T. and Bell, A. J. (2007) 'Enhancing the understanding of structural concepts – a collection of students' coursework', The University of Manchester.
11.7 Bailey, C. G., Currie, P. M. and Miller, F. R. (2006) 'Development of a new long span composite floor system', *The Structural Engineer*, Vol. 84, No. 21, pp. 32–38.
11.8 Lin, L. H., Edmonds, D. T. and Vollrath, F. (1995) 'Structural engineering of an orb-spider's web', *Nature*, Vol. 373, pp. 146–168.
11.9 Ji, T. (2001) 'Cable structures – a collection of students' coursework', UMIST.
11.10 Ji, T. (1986) 'Design and analysis of orthogonal cable-net roofs with complex shapes', *Proceedings of International Symposium on Membrane Structures and Space Frames*, Vol. 2, Japan.

12 Horizontal movements of structures induced by vertical loads

12.1 Concepts

- Vertical loads acting on structures can induce horizontal movements. Exceptions are symmetric structures subject to symmetric vertical loads and (rarely) anti-symmetric structures subject to anti-symmetric loads.
- The magnitudes of the horizontal movements of structures in response to vertical loads depend on the load distribution and the structural geometry.
- When the frequency of a dynamic vertical load is close to one of the lateral natural frequencies of a structure, resonance in the lateral direction of the structure can occur.

12.2 Theoretical background

When a structure moves horizontally, it is usually considered that this is in response to horizontal loads. However, vertical loads can also induce horizontal movements. This is because structures are three-dimensional and movements in the orthogonal directions are often coupled. For some structures such horizontal movements can be a significant design consideration, especially when dynamic response is important.

Horizontal movements may result from the following:

- Horizontal loading, for example wind loading which will generate translational movement of tall buildings.
- Loading that, although primarily vertical, has a horizontal component, for example walking. The Millennium Footbridge in London is a structure where significant horizontal movements were induced by people walking [12.1].
- Vertical loading acting on asymmetric structures. Due to the structural geometry, vertical loads can induce both vertical and horizontal movements, i.e. vertical motion is coupled with the horizontal response. An example is a simple frame that has two columns with different lengths subject to vertical loading, which will be studied in section 12.2.1.3.
- Vertical loading acting asymmetrically on structures. Due to their location, vertical loads can induce both vertical and horizontal movements. An example is that of a train crossing a bridge and producing horizontal movements orthogonal to the rails; this will be discussed later.

This chapter considers the last two situations where vertical loading can generate horizontal movements of frame structures [12.2].

140 Statics

12.2.1 Static response

12.2.1.1 A symmetric system

Consider a simple symmetric frame with no horizontal forces but subjected to a concentrated vertical load as shown in Figure 12.1a. The beam has a length of L and rigidity of EI_b and the two columns have the same length of h and rigidity of EI_c.

If the axial deformations of the columns and the beam of the frame can be considered to be negligible, the structure has three degrees of freedom; the horizontal displacement, u, and the rotations, θ_A and θ_B at the connections of the beam and columns. Thus the equations of static equilibrium of the frame are given by:

$$\frac{EI_c}{h^3}\begin{bmatrix} 24 & 6h & 6h \\ 6h & 4h^2(\alpha\beta+1) & 2h^2\alpha\beta \\ 6h & 2h^2\alpha\beta & 4h^2(\alpha\beta+1) \end{bmatrix}\begin{Bmatrix} u \\ \theta_A \\ \theta_B \end{Bmatrix} = \begin{Bmatrix} 0 \\ M_A \\ M_B \end{Bmatrix} \qquad (12.1)$$

where

$$\alpha = h/L \quad \beta = EI_b/EI_c \qquad (12.2)$$

M_A and M_B are the fixed end moments of the beam produced by the vertical loading. By convention the positive sign occurs when the end moment induces clockwise rotation. As the coefficient matrix in equation 12.1 is fully populated, the horizontal displacement is coupled with the rotations.

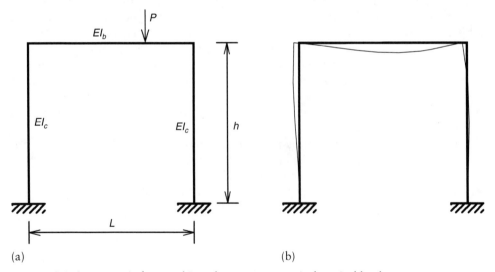

(a)   (b)

Figure 12.1 A symmetric frame subjected to an asymmetrical vertical load.

Expanding the first row of equation 12.1 gives:

$$u = -\frac{h(\theta_A + \theta_B)}{4} \qquad (12.3)$$

Therefore u is zero when $\theta_A = -\theta_B$. This occurs when symmetric loads are applied to the beam. Solving equation 12.1 gives the horizontal movement of the frame due to the vertical load:

$$u = \frac{-(M_A + M_B)}{4(6\alpha\beta + 1)\alpha L} \frac{h^3}{EI_c} \tag{12.4}$$

The negative sign indicates that the movement of the frame is to the left.

Now consider a horizontal load that will induce the same horizontal movement as the vertical load. This will be termed an equivalent horizontal load. If a horizontal force of F to the left is applied at one of the beam/column connections instead of the vertical load, the solution of equation 12.1 gives the horizontal displacement as:

$$u = \frac{-(3\alpha\beta + 2)F}{12(6\alpha\beta + 1)} \frac{h^3}{EI_c} \tag{12.5}$$

For the same horizontal displacement at the beam–column connections of the frame, equating equations 12.4 and 12.5 gives the equivalent horizontal load:

$$F = \frac{12(M_A + M_B)(6\alpha\beta + 1)}{4(6\alpha\beta + 1)(3\alpha\beta + 2)\alpha L}$$

$$= \frac{(M_A + M_B)}{LP_{TV}} \frac{3}{(3\alpha\beta + 2)\alpha} P_{TV} = C_L C_S P_{TV} = C_{LS} P_{TV} \tag{12.6}$$

in which

$$C_L = \frac{M_A + M_B}{LP_{TV}} \tag{12.7}$$

$$C_S = \frac{3}{(3\alpha\beta + 2)\alpha} \tag{12.8}$$

$$C_{LS} = C_L C_S \tag{12.9}$$

where P_{TV} is the total vertical load, for the studied case $P_{TV} = P$; C_L is a load factor that relates to the type and distribution of vertical loads; C_S is a structure factor that is a function of structural form (α) and the distribution of member rigidities (β); C_{LS} is an equivalent horizontal load factor.

It can be seen that the smaller the values of α and β the larger the structure factor C_S. It should also be noted that, for this case, the load factor and the structure factor are independent.

Equation 12.6 indicates that the equivalent horizontal load can be expressed as a product of the load factor, the structure factor and the vertical load. Table 12.1 provides values of the load factor C_L for several vertical load distributions on a beam with two fixed ends. Table 12.2 shows the structure factors C_S for a range of geometry and rigidity ratios.

142 Statics

Table 12.1 The load factor (C_L) for different load distributions on a uniform beam with two fixed ends

Load distribution	M_A	M_B	C_L
Uniformly distributed load over full length	$-qL^2/12$	$qL^2/12$	0
Concentrated load acting at a quarter of the span from the right	$-3PL/64$	$9PL/64$	3/32
Uniformly distributed load over a half of the span from right	$-5qL^2/192$	$11qL^2/192$	1/16
Uniformly distributed load over three-quarters of the span from right	$-63qL^2/1024$	$81qL^2/1024$	3/128

Table 12.2 The structure factor (C_S) for different ratios of length and rigidity for a symmetric frame

	$\beta = 0.5$	$\beta = 1$	$\beta = 2$
$\alpha = 0.5$	2.182	1.714	1.200
$\alpha = 1$	0.8570	0.6000	0.3750
$\alpha = 2$	0.3000	0.1875	0.1071

Table 12.3 The equivalent horizontal load factor (C_{LS}) for a symmetric frame

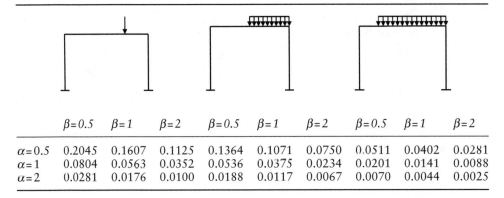

	$\beta=0.5$	$\beta=1$	$\beta=2$	$\beta=0.5$	$\beta=1$	$\beta=2$	$\beta=0.5$	$\beta=1$	$\beta=2$
$\alpha=0.5$	0.2045	0.1607	0.1125	0.1364	0.1071	0.0750	0.0511	0.0402	0.0281
$\alpha=1$	0.0804	0.0563	0.0352	0.0536	0.0375	0.0234	0.0201	0.0141	0.0088
$\alpha=2$	0.0281	0.0176	0.0100	0.0188	0.0117	0.0067	0.0070	0.0044	0.0025

Consider a particular case where $\alpha = 1$, $\beta = 1$ and a concentrated load P acts at a quarter the length of the horizontal member of the frame as shown in Figure 12.1a. The moments in equation 12.1 are:

$$M_A = -\frac{3Ph}{64} \quad M_B = \frac{9Ph}{64} \qquad (12.10)$$

Substituting these into equation 12.6 gives:

$$F = \frac{63P}{1130} = 0.05625P \qquad (12.11)$$

Horizontal movements of structures 143

In this case the horizontal movement of the frame at the beam–column connections induced by the vertical load P is equal to that of a horizontal load of 5.625 per cent of P applied on one of the connections in the direction of the movement of the frame.

Example 12.1

Consider the frame shown in Figure 12.1a with $h = 6$ m, $L = 6$ m, $E = 30 \times 10^9 \, \text{N/m}^2$, $I_b = I_c = 0.25^4/12 = 3.255 \times 10^{-4} \, \text{m}^4$ and for two loading cases:

1. A vertical concentrated load, $P = 100$ kN, acts at one-quarter of the length of the beam from the right end of the frame.
2. A horizontal concentrated load, $F = 0.05625P = 5.625$ kN, acts on the left beam–column connection towards to the left.

Calculate the horizontal displacements induced by the two loading cases respectively.

Solution

Figure 12.1b shows the deformed shape of the frame subject to the concentrated vertical load. The horizontal displacements determined by solving equation 12.5 and by using the finite element method are -7.406 mm and -7.405 mm respectively.

The lateral displaced induced by the equivalent lateral load is also -7.406 mm through the computer analysis.

The combined effect of the load and structure factors are shown in Table 12.3, which provides the equivalent horizontal load factors C_{LS} for $\alpha = 0.5$, 1 and 2, and $\beta = 0.5$, 1 and 2.

From Tables 12.1, 12.2 and 12.3 it can be seen that:

- The magnitude of the horizontal displacement induced by vertical loads (equation 12.4) or the equivalent horizontal load (equation 12.6) depends on the load distribution and the structural form.
- The structure factors C_S are significantly larger than the load factors C_L.
- The smaller the values of α and β, the larger the equivalent horizontal load factor, i.e. if the frame is relatively short and has a relatively large span, and/or the rigidity of the beam is smaller than that of the column, the frame will have a relatively large horizontal load factor. Hence, the equivalent horizontal load for a taller frame is smaller than that for a shorter frame if both are subjected to the same vertical loading.
- The height/length ratio α is more significant than the rigidity ratio β when determining the magnitude of the horizontal movement.
- The horizontal movement of a frame due to vertical loads is zero when $M_A = -M_B$, i.e. when concentrated loads act on the beam/column joints or when a symmetric load is applied to the beam.

It is useful to explain why the frame moves to its left when the vertical load is applied on the right half of the frame.

144 Statics

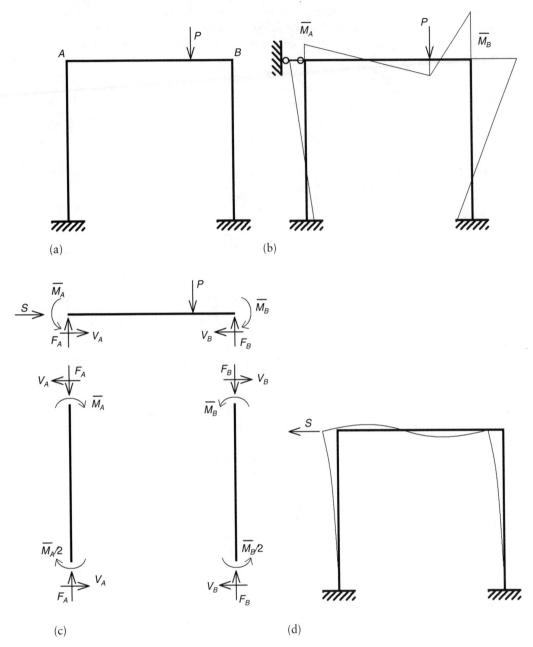

Figure 12.2 Conceptual analysis of the lateral movements of a simple frame subjected to an asymmetric vertical load.

Applying a lateral constraint at the left-hand beam–column joint, the bending moment diagram of the frame can be drawn for no lateral movements of the frame as shown in Figure 12.2b where $\overline{M}_B > \overline{M}_A$ due to the location of the load. The free-body diagram of the beam and two columns of the frame can be plotted (Figure 12.2c) based on the bending moment diagram, where V_A, F_A and V_B, F_B are the internal shear forces and axial forces on the connections A and B respectively, and S is the force in the horizontal support at A assumed to be towards the right.

Considering the equilibrium of the beam in the horizontal direction gives:

$$S + V_A - V_B = 0$$

As $M_B > M_A$, then $V_B > V_A$. Thus S is positive, i.e. acting in the assumed direction. As there is no horizontal restraint on the beam–column connection in the actual frame, the action of S must be removed. Following the principle of superposition, this can be achieved by adding a force $-S$ on the left beam–column connection as shown in Figure 12.2d. Obviously $-S$ causes the frame to move to its left. Adding the two loading conditions shown in Figures 12.2b and 12.2d together forms the actual frame with the vertical load (Figure 12.2a) and adding together the displacements of the two loading cases shows horizontal deflections of the frame.

12.2.1.2 An anti-symmetric system

If the left column of the frame shown in Figure 12.1a is rotated through 180 degrees about its connection to the beam, the frame becomes anti-symmetric as shown in Figure 12.3a. The equivalent horizontal load can be found, similar to that in section 12.2.2.1, as:

$$F = \frac{(M_B - M_A)}{4(2\alpha\beta + 1)\alpha L} \frac{12(2\alpha\beta + 1)}{(\alpha\beta + 2)}$$

$$= \frac{(M_B - M_A)}{LP_{TV}} \frac{3}{(\alpha\beta + 2)\alpha} P_{TV} = C_L C_S P_{TV} = C_{LS} P_{TV} \quad (12.12)$$

where

$$C_L = \frac{M_B - M_A}{LP_{TV}} \quad (12.13)$$

$$C_S = \frac{3}{(\alpha\beta + 2)\alpha} \quad (12.14)$$

where C_{LS} is defined by equation 12.9. Equations 12.12, 12.13 and 12.14 have the same form as equations 12.6, 12.7 and 12.8. For comparison similar tables for the load factor, the structure factor and the equivalent horizontal load factor of the anti-symmetric frame are given in Tables 12.4, 12.5 and 12.6.

In addition to the conclusions drawn from section 12.2.1.1 which are also valid for the anti-symmetric system, it can be deduced that:

- The load factor C_L and structure factor C_S for the anti-symmetric systems are significantly larger than those for the symmetric system. Hence, the magnitude of the horizontal movement due to vertical loads depends primarily on the structural form.

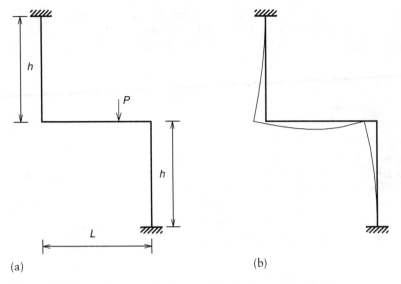

Figure 12.3 An anti-symmetric frame with an asymmetrical load.

- Equation 12.12 indicates that the anti-symmetric frame has no horizontal movement when $M_A = M_B$, which requires a particular distribution of anti-symmetric vertical loading. For any other vertical loading there will be a resulting horizontal movement.

Example 12.2

Consider the frame shown in Figure 12.3a with similar properties to those used for example 12.1, i.e. $h = 6\,\text{m}$, $L = 6\,\text{m}$, $E = 30 \times 10^9\,\text{N/m}^2$, $I_b = I_c = 0.25^4/12 = 3.255 \times 10^{-4}\,\text{m}^4$ and $P = 100\,\text{kN}$ (acting at one-quarter the length of the beam from the right end). Calculate the horizontal movement of the anti-symmetric frame at the beam–column connection.

Solution

The equivalent horizontal load can be evaluated using equation 12.12 and is 18.75 kN. Computer analysis is used to determine the horizontal displacements induced first by the vertical load of 100 kN, and then by a horizontal load of 18.75 kN, applied at the beam–column connection, towards the left. The displacements are found to be identical and have a value of −34.56 mm. Figure 12.3b shows the deformed shape of the frame subject to the concentrated vertical load.

12.2.1.3 An asymmetric system

If the lengths of the columns of the frame shown in Figure 12.1a are different, the frame becomes asymmetric, as shown in Figure 12.4a. The ratios given in equation 12.9 are redefined as follows:

$$\alpha = h_1/L \quad \beta = EI_b/EI_c \quad \gamma = h_1/h_2 \tag{12.15}$$

Horizontal movements of structures 147

Table 12.4 The load factor (C_L) for different load distributions for an anti-symmetric system

Load distribution	M_A	M_B	C_L
Uniformly distributed load over full length	$-qL^2/12$	$qL^2/12$	1/6
Concentrated load acting at a quarter of the span from the right	$-3PL/64$	$9PL/64$	3/16
Uniformly distributed load over a half of the span from right	$-5qL^2/192$	$11qL^2/192$	1/6
Uniformly distributed load over three-quarters of the span from right	$-63qL^2/1024$	$81qL^2/1024$	3/16

Table 12.5 The structure factor (C_S) for different ratios of length and rigidity for an anti-symmetric frame

	$\beta=0.5$	$\beta=1$	$\beta=2$
$\alpha=0.5$	2.667	2.400	2.000
$\alpha=1$	1.200	1.000	0.750
$\alpha=2$	0.500	0.375	0.250

Table 12.6 The equivalent horizontal load factor (C_{LS}) for an anti-symmetric frame

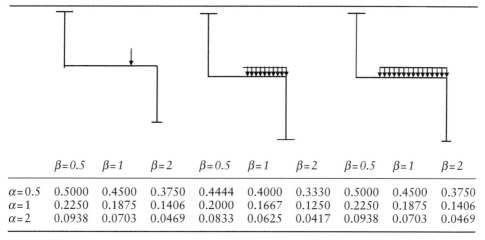

	$\beta=0.5$	$\beta=1$	$\beta=2$	$\beta=0.5$	$\beta=1$	$\beta=2$	$\beta=0.5$	$\beta=1$	$\beta=2$
$\alpha=0.5$	0.5000	0.4500	0.3750	0.4444	0.4000	0.3330	0.5000	0.4500	0.3750
$\alpha=1$	0.2250	0.1875	0.1406	0.2000	0.1667	0.1250	0.2250	0.1875	0.1406
$\alpha=2$	0.0938	0.0703	0.0469	0.0833	0.0625	0.0417	0.0938	0.0703	0.0469

and the equivalent horizontal load becomes:

$$F = \frac{3[(\alpha\beta(2-\gamma^2)+2\gamma]M_A + 3[\alpha\beta(2\gamma^2-1)+2\gamma^2]M_B}{\alpha L[4(\alpha\beta+1)\gamma+\alpha\beta(3\alpha\beta+4)]P_{TV}} P_{TV} = C_{LS}P_{TV} \quad (12.16)$$

where:

$$C_{LS} = \frac{3[(\alpha\beta(2-\gamma^2)+2\gamma]M_A + 3[\alpha\beta(2\gamma^2-1)+2\gamma^2]M_B}{\alpha L[4(\alpha\beta+1)\gamma+\alpha\beta(3\alpha\beta+4)]P_{TV}} \quad (12.17)$$

C_{LS} is an equivalent horizontal load factor, which is a function of the load distribution, location and structural form. In contrast to the symmetric and anti-symmetric frames considered in the previous two sections, the load factor and the structure factor are coupled for the asymmetric frame.

Consider $\gamma = 3/2$. The equivalent horizontal load factors for the same loading cases, length ratios α and rigidity ratios β, as for the symmetric and anti-symmetric frames, are given in Table 12.7.

Comparing the results in Tables 12.3 and 12.7, it can be seen that the equivalent horizontal load factors for the asymmetric frame are significantly larger than those for the symmetric frame. This again shows that structural form affects the magnitudes of horizontal movements of frame structures subject to vertical loads.

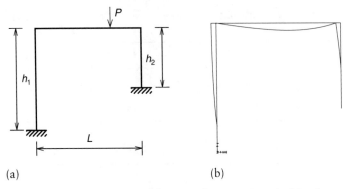

Figure 12.4 An asymmetrical frame with an asymmetrical load.

Example 12.3

Consider the frame shown in Figure 12.4a with $h_1 = L = 6.0\,\text{m}$, $h_2 = 4\,\text{m}$, $E = 30 \times 10^9\,\text{N/m}^2$, $I_b = I_c = 3.255 \times 10^4\,\text{m}^4$ and $P = 100\,\text{kN}$ (acting at one-quarter the length of the beam from the right end). Calculate the horizontal movement of the asymmetric frame at the beam–column connection.

Solution

The equivalent horizontal load for this case is evaluated using equation 12.16 and is 15.73 kN. The horizontal displacements of the frame from computer analysis for the

Table 12.7 The equivalent horizontal load factor (C_{LS}) for an asymmetric system

	$\beta=0.5$	$\beta=1$	$\beta=2$	$\beta=0.5$	$\beta=1$	$\beta=2$	$\beta=0.5$	$\beta=1$	$\beta=2$
$\alpha=0.5$	0.4269	0.3800	0.3146	0.3197	0.2892	0.2442	0.2251	0.2162	0.1952
$\alpha=1$	0.1900	0.1573	0.1184	0.1446	0.1221	0.0938	0.1081	0.0976	0.0796
$\alpha=2$	0.0786	0.0592	0.0400	0.0616	0.0469	0.0322	0.0488	0.0398	0.0285

vertical loading and a horizontal load of 15.73 kN applied to the left at the left-hand beam–column connection have the same value of −10.59 mm. Figure 12.4b shows the deformed shape of the frame.

12.2.1.4 Further comparison

Table 12.8 summarises the ranges of the equivalent horizontal load factors for the three types of frame subject to three types of vertical loading for variations of α and β between 0.5 and 2. From Table 12.8 it can be seen that:

- The equivalent horizontal load factors for the anti-symmetric frame have the largest values, but this type of structure may not be common.
- The equivalent horizontal load factors of the asymmetric frame are at least double those of the symmetric frame for the same loading conditions.

Table 12.8 Comparison of the ranges of the equivalent horizontal load factor (C_{LS})

	Symmetric frame	Anti-symmetric frame	Asymmetric frame
Concentrated load acting at one-quarter of the span from the right	0.2045–0.0100	0.5000–0.0469	0.4269–0.0400
Uniformly distributed load over a half of the span from right	0.1364–0.0067	0.4444–0.0417	0.3197–0.0322
Uniformly distributed load over three-quarters of the span from right	0.0511–0.0025	0.5000–0.0469	0.2251–0.0285

12.2.2 Dynamic response

When a structure is subjected to dynamic loading, resonance may occur with a consequent, and potentially significant, increase in response (see Chapters 15 and 16). The possibility of vertical dynamic loading resulting in a resonant horizontal response therefore must be considered.

Consider the frame discussed in 12.2.1.1 and shown in Figure 12.1a subjected to a simple sinusoidal vertical load, $P(t)$, with maximum amplitude P_0:

$$P(t) = P_0 \sin 2\pi f_p t \qquad (12.18)$$

where f_p is the frequency of the load and t is time. The mass densities for the columns and the beam are assumed to be $\overline{m}$ and $10\overline{m}$ respectively, with the high density of the beam representing added loads that may arise from floors supported by the beam, etc. The equation of undamped forced vibrations of the frame can be shown [12.2]:

$$\frac{\overline{m}h}{420} \begin{bmatrix} 4512 & 22h & 22h \\ 22h & 44h^2 & -30h^2 \\ 22h & -30h^2 & 44h^2 \end{bmatrix} \begin{Bmatrix} \ddot{u} \\ \ddot{\theta}_A \\ \ddot{\theta}_B \end{Bmatrix} + \frac{EI_c}{h^3} \begin{bmatrix} 24 & 6h & 6h \\ 6h & 8h^2 & 2h^2 \\ 6h & 2h^2 & 8h^2 \end{bmatrix} \begin{Bmatrix} u \\ \theta_A \\ \theta_B \end{Bmatrix} = \begin{Bmatrix} 0 \\ M_A \\ M_B \end{Bmatrix} \sin(2\pi f_p t)$$

$$(12.19)$$

150 *Statics*

The elements in the coefficient matrix for accelerations are obtained in the same manner as those in coefficient matrix for displacements. The mode shapes and natural frequencies of the structure can be found by solving the eigenvalue problem associated with equation 12.19. Taking the mass density, $\overline{m}$, equal to 150 kg/m and other data as used in example 12.1, the three natural frequencies of the frame are 1.39 Hz, 5 Hz and 14.5 Hz and the corresponding mode shapes are shown in Figure 12.5. The first mode shows the dominated horizontal movements of the frame while the two other modes give symmetric and anti-symmetric rotations of the two connections of the frame respectively. The response in the first mode of the frame is:

$$A_1(t) = \frac{\phi_{21}M_A + \phi_{31}M_B}{K_1} \frac{1}{1-(f_p/f_1)^2} \sin 2\pi f_p t \qquad (12.20)$$

where $A_1(t)$ is the amplitude of the horizontal motion of the frame and $\phi_{21}M_A + \phi_{31}M_B$ is the load for the first mode, which acts in the horizontal direction. Equation 12.20 indicates that if the load for the first mode is not equal to zero and the load frequency (f_p) is close to the fundamental natural frequency (f_1), the vertical load will induce near resonant vibration of the frame in the horizontal direction. This conclusion can be verified numerically.

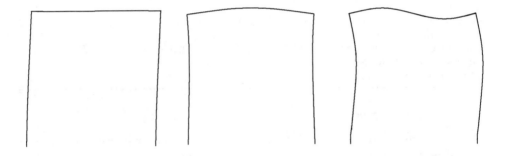

(a) Horizontal movement   (b) Symmetric movement   (c) Anti-symmetric movement

Figure 12.5 Mode shapes of the symmetric frame.

Example 12.4

Consider the frame defined in example 12.1 with the additional data as follows:

$\overline{m} = 2400 \text{ kg/m}^3 \times (0.25 \text{ m})^2 = 150 \text{ kg/m}$, $f_p = 1.39 \text{ Hz}$, $P_0 = 100 \text{ kN}$ and $P(t) = P_0 \sin 2\pi f_p t$

Calculate the dynamic displacements of the frame in the lateral direction using equation 12.19.

Solution

Dynamic analysis was carried out by computer with the damping set to zero. Figure 12.6 shows the time history of the horizontal motion of the frame, up to ten seconds, due to the vertical harmonic load. A typical resonance situation is encountered.

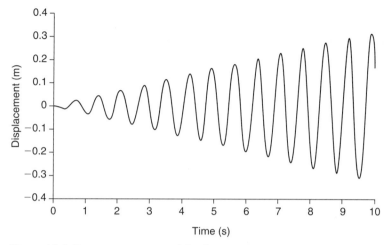

Figure 12.6 Resonant response of the frame.

Although the example is simple, it illustrates the important phenomenon that:

If the frequency of a vertical load is close to one of the horizontal natural frequencies of a structure, resonance in the horizontal direction can occur as a result of vertical excitation.

This situation should be recognised in the design of structures.

The necessary condition for no horizontal movement of the frame occurs when the vertical loads are applied either symmetrically on the beam or at the beam–column joints. For any other distributions of vertical dynamic loads, resonance can occur with motion in the horizontal direction.

12.3 Model demonstrations

12.3.1 A symmetric frame

Figure 12.7 shows a simple symmetric plastic frame unloaded (a) and carrying an asymmetrically concentrated load positioned close to the right-hand column (b). It

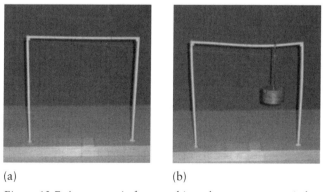

(a) (b)

Figure 12.7 A symmetric frame subjected to an asymmetric load.

152 *Statics*

can be observed that the horizontal member deflects vertically and the loaded frame moves to the left. Note that the movement is to the left for the load placed to the right of the centre line of the frame. If the vertical load was a harmonic dynamic load and its frequency matched the natural frequency of the frame in its horizontal direction, resonance with significant horizontal movements of the frame would occur.

12.3.2 An anti-symmetric frame

Figure 12.8 shows a simple anti-symmetric plastic frame unloaded (a) and carrying an asymmetrically concentrated load positioned close to the right-hand column (b). It can be observed that the horizontal member deflects vertically and the loaded frame moves to its left. Again the movement is to the left for the load placed to the right of the centre line of the frame.

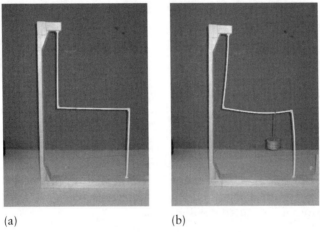

(a)   (b)

Figure 12.8 An anti-symmetric frame subjected to an asymmetric load.

12.3.3 An asymmetric frame

Figure 12.9 shows a simple asymmetric plastic frame unloaded (a) and carrying a concentrated load positioned close to the right-hand column (b). The horizontal

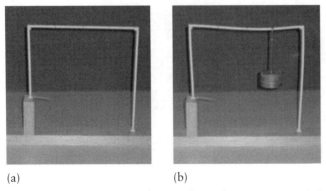

(a)   (b)

Figure 12.9 An asymmetric frame subjected to an asymmetric load.

member deflects vertically and it can be observed that the loaded frame moves to the left. Again the movement is to the left for the load placed to the right of the centre line of the frame.

12.4 Practical examples

12.4.1 A grandstand

Figure 12.10 shows coupled vertical and front-to-back movement of the cross-section of a grandstand in one typical mode of vibration. It can be seen that the front-to-back movements of the grandstand are larger than the vertical movements of the two tiers for this particular mode. This mode shape indicates that resonance in the front-to-back direction would occur if one of the frequencies of vertical loading on a tier was close to the natural frequency associated with the mode, even though the vertical movement will be small. It has been observed at a pop concert that a stand moved much more significantly in the front-to-back direction than in the vertical direction although the human loading was primarily in the vertical direction.

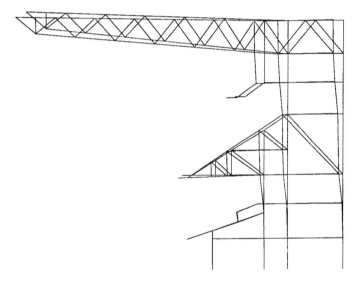

Figure 12.10 Typical mode of vibration of a frame model of a cantilever grandstand showing coupled vertical and front-to-back movements [12.3].

12.4.2 A building floor

A 9 m by 6 m test panel of a large composite floor is shown in Figure 12.11. The structural response of the panel was measured for a group of 64 students jumping following a music beat (Figure 12.12). At the centre of the test floor panel, the vertical acceleration was recorded for just over 16 s, as was the horizontal acceleration in the direction orthogonal to the direction in which the students were facing. The peak vertical acceleration was 0.48 g and the corresponding horizontal acceleration

0.03 g. The autospectra for these records are shown in Figure 12.13 and the characteristic response can be seen in both directions. The test area was part of the much larger flooring system (Figure 12.11) and the vertical human loading was thus applied asymmetrically on the structure as a whole, which induced the horizontal motion of the whole building system.

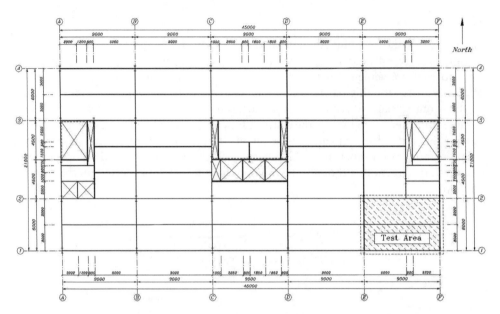

Figure 12.11 A plan of a floor used for crowd jumping tests at its corner panel.

Figure 12.12 64 students jumping on a floor in response to music.

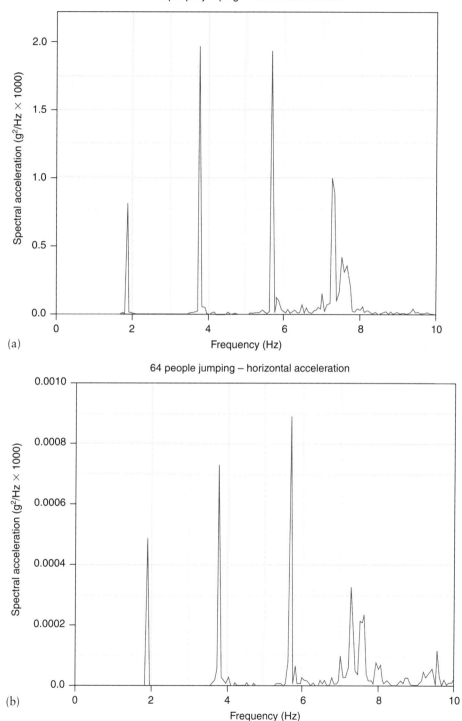

Figure 12.13 The acceleration spectrum in the vertical and horizontal directions for 64 students jumping on a floor.

12.4.3 Rail bridges

Horizontal movements of some railway bridges have been observed as trains passed over them. Because of the increasing speed of trains a number of bridges have had to be re-assessed for safety. As there are often two or more rail tracks on a bridge, the loading from any one train is effectively applied in an asymmetrical manner on the structure generating lateral horizontal movements of the bridge. There will also be some horizontal forces generated by lateral movement of the railway vehicles, even along straight tracks. With the increasing speed of trains, the vertical loading frequency will increase and this may be a problem if resonance occurs, i.e. if one of the train load frequencies in the vertical direction is close to one of the lateral natural frequencies of the bridge. It is therefore necessary to check horizontal as well as vertical natural frequencies of bridges to ensure that both are above the likely loading frequencies associated with trains running at higher speeds.

References

12.1 Dallard, P., Fitzpatrick, T., Flint, A., Low, A., Smith, R. R. and Willford, M. (2000) 'Pedestrian induced vibration of footbridges', *The Structural Engineer*, Vol. 78, No. 23/24, pp. 13–15.

12.2 Ji, T., Ellis, B. R. and Bell, A. J. (2003) 'Horizontal movements of frame structures induced by vertical loads', *Structures and Buildings: The Proceedings of the Institution of Civil Engineers*, Vol. 156, No. 2, pp. 141–150.

12.3 Mandal, P. and Ji, T. (2004) 'Modelling dynamic behaviour of a cantilever grandstand', *Structures and Buildings: The Proceedings of the Institution of Civil Engineers*, Vol. 157, No. 3, pp. 173–184.

Part II
Dynamics

13 Energy exchange

13.1 Definitions and concepts

Conservative systems: a system is said to be conservative if no energy is added or lost from the system during movement. This is an idealised system but one that can be used to aid the solution of many problems. In real structures internal friction forces will do work and damping will dissipate energy.

Conservation of energy means that the total energy at two different positions or at two different times is the same in a conservative system.

Conservation of momentum indicates that the momentum of a system is the same at two different times when the resulting external force is zero between those two times.

- Energy can be transformed from one form to another, for instance, mechanical energy can be changed into electrical energy.
- For a conservative system, the total energy is constant and a body once moved will continue to move or to vibrate. During motion there is a constant exchange between potential energy and kinetic energy.
- For a non-conservative system, energy has to be added continuously to maintain motion.

13.2 Theoretical background

Gravitational potential energy of a mass is defined as the work done against gravity to elevate the mass a distance above an arbitrary reference position where the gravitational potential energy is defined to be zero. Thus the potential energy is:

$$U_g = mgh \tag{13.1}$$

where m is the mass of the body, g is the gravitational acceleration and h is the height of the mass above the reference position.

Elastic potential energy is the potential energy found in the deformation of an elastic body, such as a spring or a deformable beam. The elastic potential energy in a spring with stiffness k when it is displaced with a distance of u is defined as:

$$U_e = \frac{1}{2} ku^2 \tag{13.2}$$

160 Dynamics

This is the total work required to take the mass from its original position to a displacement of u. For a deformable beam with rigidity EI, the elastic potential energy due to bending of the beam is expressed as:

$$U_e = \frac{1}{2}\int_0^L EI\left(\frac{d^2v}{dx^2}\right)^2 dx \qquad (13.3)$$

where $v(x)$ is the deformation of the beam along its length. The elastic potential energy defined by equation 13.3 is also called the strain energy of the beam. The strain energy due to shear forces can be disregarded if the length of a beam is much greater than the depth (say the ratio of the span to the depth of the beam is larger than 8) [13.1]. Strain energy or elastic potential energy is always positive, regardless of the direction of the displacement.

Kinetic energy of a mass of m with a velocity of $\dot{u}$ is defined as:

$$T = \frac{1}{2}m\dot{u}^2 \qquad (13.4)$$

This is equal to the work required to move the mass from a state of rest to a velocity of $\dot{u}$. As is the case for the strain energy, the kinetic energy is always positive regardless of the direction of motion. For a vibrating beam with distributed mass of $m(x)$, the kinetic energy is:

$$U_e = \frac{1}{2}\int m(x)\dot{v}^2 dx \qquad (13.5)$$

All the forms of energy are scalar quantities with SI units of Nm or Joules (J).

Conservative systems: for a conservative system, the total energy is constant. In other words, if the energy of the system is calculated at two different locations or two different times, the values of energy at the two locations/times are the same, i.e.:

$$U_1 = U_2 \quad \text{or} \quad U_1 - U_2 = 0 \qquad (13.6)$$

Equation 13.6 shows the *principle of conservation of energy* and indicates that *energy can be transformed from one form to another whilst keeping the total energy constant*. The principle of conservation of energy (equation 13.6) leads to a useful method of analysis.

The basic equation of motion of a mass m moving in a particular direction is:

$$\sum F = m\ddot{u} = \frac{d}{dt}(m\dot{v}) \quad \text{or} \quad \sum F = \frac{d}{dt}(G) = \dot{G} \qquad (13.7)$$

where F is the force acting on the mass and $\ddot{u}$ is the acceleration of the mass. The product of the mass and velocity is defined as the momentum $G = m\dot{v}$. Equation 13.7 indicates that the resultant of all forces acting on the mass equals its rate of the change of momentum with respect to time. Momentum has SI unit of kg m/s or Ns.

Equation 13.7 can be extended to other directions, including rotation. If the resultant force on a mass or a system is zero during an interval of time, the momentum in that interval is constant, i.e.:

$$G_1 = G_2 \quad \text{or} \quad G_1 - G_2 = 0 \tag{13.8}$$

Equation 13.8 shows the *principle of conservation of momentum* and states that *the momentum at a time interval is constant if the resultant force is zero during that time interval*. This principle of the conservation of momentum (equation 13.8) also leads to a method for solving some problems. In addition, equation 13.8 can be applied to both conservative and non-conservative systems.

Example 13.1

A 5 kg cylinder is released from rest in the position shown in Figure 13.1 and compresses a spring of stiffness $k = 1.8$ kN/m. Determine the maximum compression v_{max} of the spring and the maximum velocity $\dot{v}_{max}$ of the cylinder [13.1].

Solution

Consider the system to be conservative in which the effect of air resistance and any friction is negligible. The cylinder moves down v_{max} before it rebounds due to the spring action.

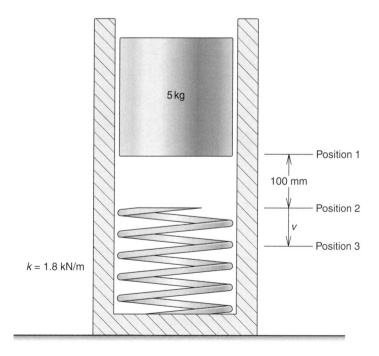

Figure 13.1 [13.1] (permission of John Wiley & Sons Inc.).

Before releasing the cylinder from the rest (position 1), the total gravitational potential and elastic potential energies are:

$$U_{g1} = mgh = 5 \times 9.81 \times (0.1 + v_{max}) \quad \text{and} \quad U_{e1} = 0$$

The reference position is selected to be where the spring experiences its maximum deflection. The total energies at the reference position (position 2) are:

$$U_{g2} = 0 \quad \text{and} \quad U_{e2} = \frac{1}{2}kv_{max}^2 = 900v_{max}^2$$

Using the condition of energy conservation (equation 13.6) leads to:

$$900v_{max}^2 = 5 \times 9.81(0.1 + v_{max}) \quad \text{or} \quad 900v_{max}^2 - 49.5v_{max} - 49.5 = 0$$

Solving the above equation gives $v_{max} = 0.264$ m.

The total gravitational potential and kinetic energies at the position immediately before the cylinder contacts the spring (position 3) are:

$$U_{g3} = mgh = 5 \times 9.81 \times 0.1 = 4.95 \, \text{Nm} \quad \text{and} \quad T_3 = \frac{1}{2}m\dot{v}_{max}^2 = 2.5\dot{v}_{max}^2$$

Equating the total energy at positions 1 and 3 gives:

$$4.95 + 2.5\dot{v}_{max}^2 = 5 \times 9.81(0.1 + 0.264) = 17.9$$

Thus the maximum velocity of the cylinder is:

$$\dot{v}_{max} = \sqrt{17.9 - 4.95} = 3.59 \, \text{m/s}$$

Example 13.2

Consider a simply supported beam of length L, with flexural rigidity of EI and a uniformly distributed mass of $\overline{m}$. The vibration in the fundamental mode of the beam is defined as:

$$v(x,t) = A_0 \sin\omega t \sin\frac{\pi x}{L} \tag{13.9}$$

where A_0 is the vibration amplitude and ω is the natural frequency of the vibration. Determine the expression for the natural frequency of the beam.

Energy exchange 163

Solution

The total energy of the system at anytime can be calculated, but the simplest case to consider is when the strain energy reaches its maximum while the kinetic energy is zero, or when the kinetic energy reaches its maximum while the strain energy is zero.

At $t = \pi/(2\omega)$, the displacement and velocity of the beam are:

$$v\left(x, \frac{\pi}{2\omega}\right) = A_0 \sin\omega\left(\frac{\pi}{2\omega}\right)\sin\frac{\pi x}{L} = A_0 \sin\frac{\pi x}{L}$$

$$\dot{v}\left(x, \frac{\pi}{2\omega}\right) = A_0 \omega \cos\omega\left(\frac{\pi}{2\omega}\right)\sin\frac{\pi x}{L} = 0$$

Thus the total energy is equal to the maximum strain energy. Using equation 13.3 gives:

$$U_{e,max} = \frac{1}{2}\int_0^L EI\left(\frac{d^2v}{dx^2}\right)^2 dx = \frac{1}{2}\int_0^L EI\left(-A_0 \frac{\pi^2}{L^2}\sin\frac{\pi x}{L}\right)^2 = \frac{A_0^2 EI\pi^4}{2L^4}\frac{L}{2} = \frac{A_0^2 EI\pi^4}{4L^3}$$

At $t = \pi/\omega$, the displacement and velocity of the beam are:

$$v(x, \pi/\omega) = A_0 \sin\omega\left(\frac{\pi}{\omega}\right)\sin\frac{\pi x}{L} = 0$$

$$\dot{v}(x, \pi/\omega) = A_0 \omega \cos\omega\left(\frac{\pi}{\omega}\right)\sin\frac{\pi x}{L} = -A_0 \omega \sin\frac{\pi x}{L}$$

The total energy of the system is equal to the maximum kinetic energy. Using equation 13.5 gives:

$$T_{max} = \frac{1}{2}\int_0^L \overline{m}\dot{v}^2 dx = \frac{1}{2}\int_0^L \overline{m}\left(-A_0\omega\sin\frac{\pi x}{L}\right)^2 = \frac{A_0^2\omega^2\overline{m}}{2}\frac{L}{2} = \frac{A_0^2\omega^2\overline{m}L}{4}$$

Using equation 13.6, i.e. equating the maximum strain energy to the maximum kinetic energy, gives:

$$\omega^2 = \frac{EI\pi^4}{\overline{m}L^4}$$

A more powerful way of using energy concepts for solving problems is to use the Lagrange equation. For free vibration, this may be written as [13.2, 13.3, 13.4]:

$$\frac{d}{dt}\left(\frac{\partial T}{\partial \dot{v}}\right) + \frac{\partial U}{\partial v} = 0 \qquad (13.10)$$

where T and U are the kinetic and potential energies of the system respectively. Example 13.2 is re-analysed to show the usefulness of the Lagrange method.

The motion of the beam (13.9) is written as:

$$v(x, t) = A(t)\phi(x) = A(t)\sin\frac{\pi x}{L} \tag{13.11}$$

The kinetic energy of the system is:

$$T = \frac{1}{2}\int_0^L \overline{m}v^2 dx = \frac{1}{2}\int_0^L \overline{m}\left(\dot{A}(t)\sin\frac{\pi x}{L}\right)^2 dx = \frac{\dot{A}^2(t)\overline{m}}{2}\frac{L}{2} = \frac{\dot{A}^2(t)\overline{m}L}{4} \tag{13.12}$$

The elastic potential energy is:

$$U = \frac{1}{2}\int_0^L EI\left(\frac{d^2v}{dx^2}\right)^2 dx = \frac{1}{2}\int_0^L EI\left(-A(t)\frac{\pi^2}{L^2}\sin\frac{\pi x}{L}\right)^2 = \frac{A^2 EI\pi^4}{2L^4}\frac{L}{2} = \frac{A^2(t)EI\pi^4}{4L^3} \tag{13.13}$$

Substituting the kinetic and elastic energies into equation 13.10 leads to:

$$\ddot{A}(t) + \frac{EI\pi^4}{\overline{m}L^4}A(t) = 0 \tag{13.14}$$

Substituting $A(t) = A_0\sin(\omega t)$ into equation 13.14 gives:

$$\omega^2 = \frac{EI\pi^4}{\overline{m}L^4}$$

This is the same as that obtained in example 13.2.

Comparing the solution procedures using the principle of the conservation of energy and the Lagrange equation, it can be noted that:

- The Lagrange equation leads to an equation of motion and the natural frequency can be obtained directly from the definition, while the principle of the conservation of energy leads to the solution of the natural frequency without producing the equation of motion.
- For the Lagrange equation the kinetic and elastic potential energies only need to be represented at an arbitrary position. Using the principle of the conservation of energy requires evaluating the kinetic and elastic potential energies at two different positions or times.

13.3 Model demonstrations

13.3.1 A moving wheel

This demonstration shows *the energy exchange between several common forms of energy, including gravitational potential energy, kinetic energy, energy loss due to friction and electromagnetic energy.*

Two parallel, parabolic plastic tracks supporting an aluminium wheel are shown in Figure 13.2. Three small magnets are placed on the perimeter of the aluminium wheel. Four batteries are located in the track support and an electromagnetic field is also provided in the base support.

Figure 13.2 A moving wheel showing energy transformation.

Place the wheel at one end of the tracks and release it. The wheel rolls towards the middle of the tracks with increasing speed as the potential energy of the wheel changes to kinetic energy. When passing the middle part of the tracks, the wheel speeds up due to a charge of electromagnetic energy which is sufficient to compensate for the energy loss due to the friction between the wheel and the tracks. The wheel then moves upwards and the kinetic energy is converted to potential energy. When all the kinetic energy is converted, the wheel stops and then starts to roll back down the track for a new cycle of movement. The wheel will continue to roll backwards and forwards along the track as long as sufficient electromagnetic energy is provided to compensate for the energy loss due to friction.

13.3.2 Collision balls

This demonstration shows *the use of the principles of conservation of energy and conservation of momentum.*

Figure 13.3 shows a Newton's cradle which consists of five identical stainless steel balls suspended from above and arranged in a row with each ball just touching its neighbour(s). When one pulls a ball back and releases it, it collides with the row of the remaining balls, ejecting one at the far end. If one pulls two balls and releases them, they eject two balls at the far end as shown in Figure 13.3. Lifting and releasing three balls ejects three balls at the far end, and pulling four ejects four.

First consider the collision between two identical balls of mass m, one moves with initial velocity $\dot{u}_{1i}$ and the other is at rest with $\dot{u}_{2i} = 0$. After the collision, the two

Figure 13.3 Collision balls.

balls move with velocities $\dot{u}_{1f}$ and $\dot{u}_{2f}$ respectively. Conservation of momentum requires:

$$m\dot{u}_{1i} = m\dot{u}_{1f} + m\dot{u}_{2f} \quad \text{or} \quad \dot{u}_{1i} = \dot{u}_{1f} + \dot{u}_{2f}$$

Conservation of energy requires:

$$\frac{1}{2}m\dot{u}_{1i}^2 = \frac{1}{2}m\dot{u}_{1f}^2 + \frac{1}{2}m\dot{u}_{2f}^2 \quad \text{or} \quad \dot{u}_{1i}^2 = \dot{u}_{1f}^2 + \dot{u}_{2f}^2$$

Squaring the first equation and subtracting the second equation leads to:

$$(\dot{u}_{1f} + \dot{u}_{2f})^2 - (\dot{u}_{1f}^2 + \dot{u}_{2f}^2) = 0 \quad \text{or} \quad \dot{u}_{1f}\dot{u}_{2f} = 0$$

As $\dot{u}_{2f} \neq 0$, $\dot{u}_{1f} = 0$. This shows that after collision the first ball will be at rest [13.5, 13.6].

Now consider the case of several balls. Let m_1 be the total mass of the balls launched with velocity $\dot{u}_{1i}$ and m_2 be the total mass of the balls ejected with velocity $\dot{u}_{2f}$. Noting that after collision, m_1 becomes stationary, i.e. $\dot{u}_{1f} = 0$ and using conservation of momentum and conservation of energy gives:

$$m_1 \dot{u}_{1i} = m_2 \dot{u}_{2f}$$

$$\frac{1}{2}m_1 \dot{u}_{1i}^2 = \frac{1}{2}m_2 \dot{u}_{2f}^2$$

Squaring the first equation and subtracting $2m_1$ times the second equation gives:

$$(m_2 - m_1)m_2 \dot{u}_{2f}^2 = 0$$

This equation shows that the number of balls ejected is equal to the number of balls launched if all the steel balls are the same size, as shown in Figure 13.3.

Energy exchange   167

13.3.3 Dropping a series of balls

This model gives *an entertaining demonstration of conservation of energy.*

Figure 13.4 shows four rubber balls of different sizes placed together using a plastic bar passing through their centres with the smaller balls resting on the larger balls.

When the balls are lifted vertically for a distance of about 0.15 m from a desk and then released, the smallest ball rebounds to hit the ceiling which is over two metres above the desk. This observation can be explained using either the principle of conservation of energy or the principle of conservation of momentum.

The total mass of the four balls is 120 g and the mass of the smallest ball is 5 g. The gravitational potential energy of the balls before dropping them is estimated to be:

$$mgh = 0.12 \times 9.81 \times 0.15 = 0.177\,\text{Nm}$$

After the impact of the largest ball on the desk, the three larger balls bounce up almost together about 0.03 m. According to the conservation of energy, the total gravitational potential energy before and after impact should be the same if no energy loss takes place, i.e.:

$$0.177 = 0.115 \times 9.81 \times 0.03 + 0.005 \times 9.81 \times h_1$$

From the above equation, the predicted bounce height of the smallest ball is 2.92 m. Actually there is some loss of energy due to the impacts between the largest ball and

Figure 13.4 Dropping a series of balls.

168 *Dynamics*

the desk and between the adjacent balls, so the smallest ball would not bounce quite as high as 2.92 m.

This example can also be analysed using the principle of conservation of momentum.

13.4 Practical examples

13.4.1 Rollercoasters

Figure 13.5a shows a rollercoaster which is raised from ground level to the top of the first tower using mechanical energy. This builds up a reservoir of potential energy in the rollercoaster.

When the rollercoaster is released at the top of the tower, it moves forward and down, the potential energy being quickly converted to kinetic energy and the speed of the rollercoaster increasing accordingly. The speed reaches its maximum when the rollercoaster reaches the lowest point between two adjacent towers. The rollercoaster then moves up to the next tower, the kinetic energy changing back to potential energy, and the speed of the rollercoaster reduces as its potential energy increases. The rest of the towers, dips, twists and turns of the ride, serve to change the energy of the rollercoaster back and forth between potential energy and kinetic energy (Figure 13.5b).

During the motion, some energy will be lost due to friction and, for this reason, the first tower must be higher than all the other towers so that sufficient energy is provided to overcome the energy loss.

(a) The first tower (b) Twists and turns

Figure 13.5 Rollercoasters.

13.4.2 A torch without a battery

Torches are normally powered by batteries. However, Figure 13.6 shows an environmentally friendly torch which works on the principle of energy exchange.

When one shakes the torch, the strong magnet at the far right in the body of the torch (Figure 13.6) moves and passes through electrical wires backwards and forwards and produces an electric current. Part of the kinetic energy generated by the magnet is converted to electric energy. The electric energy then changes into chemical energy through the electronic circuit and is stored in an internal storage cell. When a user switches on the torch, the chemical energy in the storage cell converts to electric energy and the electric energy changes to light energy in the bulb.

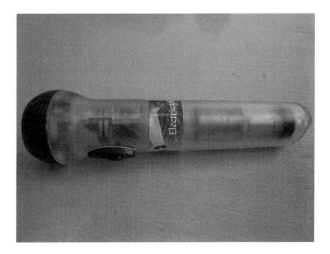

Figure 13.6 An environmentally friendly torch.

References

13.1 Meriam, J. L. and Kraige, L. G. (1998) *Engineering Mechanics: Dynamics*, Fourth Edition, New York: John Wiley & Sons.
13.2 Beards, C. F. (1996) *Structural Vibration: Analysis and Damping*, London: Arnold.
13.3 Thomson, W. T. (1966) *Theory of Vibration and Applications*, London: Allan and Unwin.
13.4 Wang, G. (1981) *Applied Analytical Dynamics*, Beijing: High Education Press (in Chinese).
13.5 Sprott, J. C. (2006) *Physics Demonstrations: A Sourcebook for Teachers of Physics*, Wisconsin: The University of Wisconsin Press.
13.6 Ehrlich, R. (1990) *Turning the World Inside Out and 174 Other Simple Physics Demonstrations*, Princeton: Princeton University Press.

14 Pendulums

14.1 Definitions and concepts

A simple gravity pendulum consists of a massless string with one end attached to a weight and the other end fixed. When an initial push is given, the pendulum will swing back and forth under the influence of gravity.

A rotational suspended system: a rigid body is suspended by two massless strings/hangers of equal length, fixed at the same point and the body rotates about the fixed point in the plane of the strings.

A translational suspended system: a rigid body is suspended by two parallel and vertical massless strings/hangers with equal length and the strings rotate about their own fixed points. Thus the rigid body moves in parallel to its static position in the plane of the strings.

- The natural frequency of a simple pendulum is independent of its mass and only relates to the length of the massless string.
- The natural frequency of a translational suspended system is independent of its mass and the location of its centre of mass and is the same as that of an equivalent simple pendulum.
- The natural frequency of a rotational suspended system is dependent on the location of its centre of mass but is independent of the magnitude of its mass.
- An outward inclined suspended system is a mechanism. When it is loaded vertically and asymmetrically, it will move sideways and rotate.
- The lateral natural frequency of a suspended bridge estimated using simple beam theory will be smaller than the true lateral natural frequency.

14.2 Theoretical background

14.2.1 A simple pendulum

Well-known for its use as a timing device, the simple pendulum is a superb tool for teaching science and engineering. It can serve as a model for the study of the linear oscillator [14.1].

Consider a pendulum consisting of a concentrated mass m suspended from a pivotal point by an inextensible string of length l as shown in Figure 14.1. Several assumptions are used to investigate the linear oscillation of a pendulum, the principal ones being:

- Negligible friction, i.e. the resulting torque on the system about the pivot is due solely to the weight of the mass.
- Small amplitude oscillations which means that the sine of the rotation, θ, can be replaced by the rotation when specified in radians.
- The mass of the string is negligible and the mass of the body is concentrated at a point.

The equation of motion of the pendulum can be established using Newton's second law directly or using the Lagrange equation. The moment induced by the weight of the mass about the pivot is $mgl\sin\theta$ where g is the acceleration due to gravity. The moment of inertia of the mass about the pivot is ml^2 and the angular acceleration is $d^2\theta/dt^2$. Using Newton's second law gives:

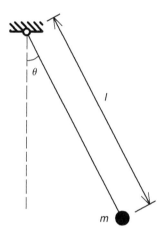

Figure 14.1 A simple pendulum.

$$ml^2 \frac{d^2\theta}{dt^2} = -mgl\sin\theta \quad \text{or} \quad \frac{d^2\theta}{dt^2} + \frac{g}{l}\sin\theta = 0 \tag{14.1}$$

Using the second assumption, $\sin\theta = \theta$ when θ is small and measured in radians. This leads to the equation of motion for small amplitude oscillations:

$$\frac{d^2\theta}{dt^2} + \frac{g}{l}\theta = 0 \tag{14.2}$$

Substituting $\theta(t) = \theta_0 \sin(\omega t)$ into equation 14.2 (or according to the definition given in Chapter 15) gives the expression for the natural frequency of the pendulum system as:

$$f = \frac{\omega}{2\pi} = \frac{1}{2\pi}\sqrt{\frac{g}{l}} \tag{14.3}$$

Equation 14.3 indicates that *the natural frequency of the pendulum system is a function of the length l and is independent of the mass of the weight.*

14.2.2 A generalised suspended system

Consider the generalised suspended system shown in Figure 14.2. This consists of a uniform rigid body symmetrically suspended by two massless and inextensible inclined strings/hangers. The top ends of the hangers are restrained by vertical and horizontal springs with stiffnesses k_x and k_y respectively. The centroid of the rigid body is constrained by horizontal, vertical and rotational springs with stiffnesses of k_{ox}, k_{oy} and k_{or} respectively.

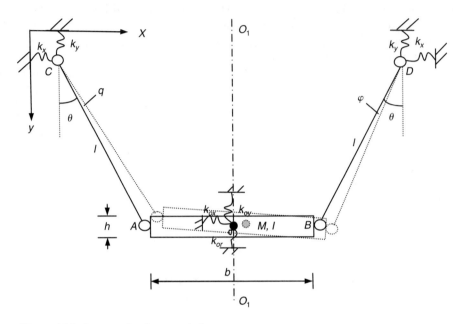

Figure 14.2 A generalised suspended system.

The rigid body has a length b, and thickness h, mass M and moment of inertia I about its centroid. The two symmetrically inclined strings have the same length l. The lower end of each string has a pinned connection to one end of the rigid body and the upper end has a pinned connection to an elastic support. The two inclined angles of the strings under gravity loading when in static equilibrium are the same and are θ. Changing the angles allows the strings to be inclined outwards ($\theta > 0$, as shown in Figure 14.2), inwards ($\theta < 0$) or vertical ($\theta = 0$). These properties form the basis for this study.

Small amplitude vibrations of the system are considered so that linearisation of the movement of the system is reasonable. It is significant to note that:

- The small second order quantities of the displacement should be considered when gravitational potential is concerned.
- Small amplitude vibrations of the system take place about its position of static equilibrium.
- The sway angle of the left string can be different to that of the right string during vibration.

This model can be used, for example, to represent a section of a suspension bridge and then used to study its lateral and torsional movements. Here the rigid body represents the bridge deck; k_x and k_y represent the constraints from the suspension cables while k_{ox}, k_{oy} and k_{or} represent the actions on the deck section from its neighbouring elements.

Due to symmetry, the model shown in Figure 14.2 has one symmetrical and two anti-symmetrical vibration modes, which can be studied separately. A detailed investigation of the dynamic characteristics of the system is given in [14.2]. The findings of some simple special cases are provided in the following subsections.

14.2.2.1 Symmetric (vertical) vibration

When $\theta = 0$ or $k_x = \infty$, as shown in Figures 14.3a and 14.3b respectively:

$$\omega^2 = \frac{k_{oy} + 2k_y}{M} = \omega_{oy}^2 + \omega_y^2 \tag{14.4}$$

where $\omega_{oy}^2 = k_{oy}/M$ and $\omega_y^2 = 2k_y/M$. This is the natural frequency of a single degree-of-freedom system consisting of a mass M and two parallel springs with stiffnesses k_{oy} and $2k_y$.

When $k_y = \infty$ as shown in Figure 14.3c, there is:

$$\omega^2 = \frac{k_{oy} + 2k_x/\tan^2\theta}{M} + \frac{\cos\theta}{\sin^2\theta}\frac{g}{l} = \omega_{oy}^2 + \frac{1}{\tan^2\theta}\omega_x^2 + \frac{\cos\theta}{\sin^2\theta}\omega_g^2 \tag{14.5}$$

in which $\omega_x^2 = 2k_x/M$ and $\omega_g^2 = g/l$. $2k_x/\tan^2\theta$ is the projected vertical stiffness of the two lateral springs due to the inclined strings/hangers and the third term in equation 14.5 shows the suspension effect.

It can be observed from equations 14.4 and 14.5 that:

- When $\theta = 0$ (Figure 14.3a), i.e. the strings/hangers are vertical, the horizontal stiffness at the hanging points has no effect on the symmetric vertical vibration of the rigid body.
- When $k_x = \infty$ (Figure 14.3b), the inclined angle has no effect on the symmetric vertical vibration.
- When $k_y = \infty$ (Figure 14.3c), both the horizontal stiffness k_x at the hanging points and gravity affect the natural frequency of the symmetric vertical vibration.
- When $k_y = \infty$, the natural frequency monotonically increases with the decreases of the inclined angle and tends to infinity when θ is close to zero.
- When $k_y = \infty$, the natural frequency is the same for $-\theta$ and θ.

14.2.2.2 Anti-symmetric (lateral and rotational) vibration

When $\theta = 0$ as shown in Figure 14.3a, the natural frequencies of the rotational and swaying vibrations of the suspended system are respectively:

$$\omega^2 = \frac{b^2 k_y}{2I} + \omega_{or}^2 \tag{14.6}$$

174 Dynamics

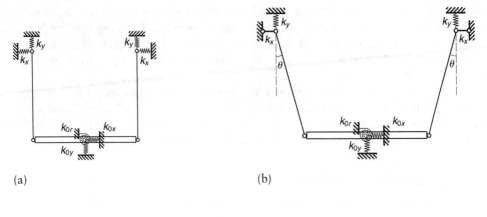

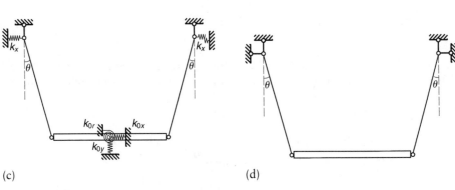

Figure 14.3 Typical suspended systems.

$$\omega^2 = \omega_{ox}^2 + \frac{\omega_g^2 \omega_x^2}{\omega_x^2 + \omega_g^2} \tag{14.7}$$

where $\omega_{or} = \sqrt{k_{or}/I}$ and $\omega_{ox} = \sqrt{k_{ox}/M}$ are the natural frequencies of the rigid body in the rotational and horizontal directions without the actions of the hangers.

It can be seen from equations 14.6 and 14.7 that:

- The rotational vibration of the rigid body is independent of the horizontal stiffnesses k_x, k_{ox} and the vertical stiffness k_{oy}.
- The swaying vibration is independent of the vertical stiffnesses k_y, k_{oy} and the rotational stiffness k_{or}.

Equation 14.7 can be rewritten as:

$$\omega^2 = \omega_{ox} \beta \tag{14.8}$$

where

$$\beta = \sqrt{1 + \frac{1}{(\omega_{ox}/\omega_g)^2 + k_{ox}/(2k_x)}} \tag{14.9}$$

Table 14.1 Values of β for varying values of ω_{ox}/ω_g and k_{ox}/k_x

ω_{ox}/ω_g	$k_{ox}/k_x = 1$	$k_{ox}/k_x = 5$	$k_{ox}/k_x = 25$	$k_{ox}/k_x = 100$
0.01	1.732	1.183	1.039	1.010
0.1	1.721	1.183	1.039	1.010
1	1.291	1.134	1.036	1.010
10	1.005	1.005	1.004	1.003
100	1	1	1	1

β is the magnification factor for the suspension effect. It can be seen from equations 14.8 and 14.9 that the lateral natural frequency of the system with suspension is always larger than that of the system without suspension. This indicates that *the lateral natural frequency of a suspended bridge estimated using simple beam theory will be smaller than the true lateral natural frequency*. Table 14.1 gives the magnification factors when ω_{ox}/ω_g varies between 0.01 and 100 and k_{ox}/k_x changes between 1 and 100. The results indicate that the larger the value of ω_{ox}/ω_g and/or the larger the value of k_{ox}/k_x, the smaller the suspension effect.

When $k_y = \infty$, $k_x = \infty$ and $k_{ox} = k_{oy} = k_{or} = 0$ as shown in Figure 14.3d, only the swaying vibration occurs. Considering a uniform rectangular rigid body, its moment of inertia about the centroid of the body is $I = M(b^2 + h^2)/12$. The solution is:

$$\omega = \omega_g \mu \tag{14.10}$$

where

$$\mu = \sqrt{\frac{1 + \tan^2\theta(1 + 2l\sin\theta/b)}{\cos\theta[1 + \tan^2\theta(1 + h/b)/3]}} \tag{14.11}$$

where μ is the magnification factor related to the inclined angle and the other design parameters. As the denominator in the square root is always positive, equation 14.11 requires a valid solution:

$$1 + \tan^2\theta(1 + 2l\sin\theta/b) > 0 \tag{14.12}$$

From equations 14.11 and 14.12 and parametric analysis [14.2] it can be seen that:

- When $\theta > 0$ the natural frequency monotonically increases with increasing b/l. However, when $\theta < 0$ the natural frequency monotonically decreases with decreasing b/l.
- The natural frequency is independent of the mass of the rigid body. Increasing the thickness–length ratio h/b results in a decrease of the natural frequency. However, for relatively small inclined angles such as $|\theta| < 40°$, the effect of h/l on the natural frequency is very small.
- The critical angle is $\theta = -\arcsin(\sqrt{b/2l})$ which can be obtained from equation 14.12. Only when the inclined angle is larger than the critical angle will the system oscillate. Otherwise, the system is in a state of unstable equilibrium, as will be demonstrated in section 14.3.1.
- The critical angle must be negative. When $b < l$, there is always a critical angle at

which the natural frequency of the system becomes zero. However, when $b \geq 2l$, equation 14.12 holds. Therefore, the necessary condition for an unstable equilibrium state is that the two strings cross each other.

14.2.3 Translational and rotational suspended systems

When $\theta = 0$ the system shown in Figure 14.3d becomes a translational suspended system in which the motion of the rigid body is around the two points where the two strings are fixed. Any position of the rigid body is therefore parallel to its original position.

In this case the magnification factor μ is equal to unity and equation 14.8 reduces to equation 14.1, i.e. the natural frequency of the translational suspended system can be calculated using the equation for a simple pendulum.

One interesting feature of the system is that the system has a constant natural frequency defined by equation 14.1 even if a mass is added to the rigid body. The size of the mass and the location of the centre of mass do not affect the natural frequency of the system.

When the two fixed points of the two strings meet, it becomes a rotational suspended system in which the motion of the rigid body is around the single fixed point. Its natural frequency can be calculated using equation 14.10. The rotational suspended system has some different dynamic characteristics from the translational suspended system.

The natural frequencies of the translational and rotational suspended systems and the effect of the added mass on the two systems will be examined through demonstration models in section 14.3.2.

14.3 Model demonstrations

14.3.1 Natural frequency of suspended systems

This model demonstration *verifies equation 14.10 and shows the unstable equilibrium state at which the suspended system will not oscillate.*

A suspended system consists of a uniform hollow aluminium bar with a square section (a rigid body) and two symmetric strings to suspend the bar. The bar has a length of $b = 0.45$ m, total mass of $M = 0.094$ kg and moment of inertia about its centroid of $I = 0.0016$ kg.m². The length and the angles of the strings can be varied.

Three typical suspension forms include vertical (Figure 14.4a, $\theta = 0$), outward inclined (Figure 14.4b, $\theta > 0$) and inward inclined (Figure 14.4c, $\theta < 0$) suspension systems. In the tests, an initial lateral displacement is applied to the bar and the bar is suddenly released to generate free vibrations. The number of oscillations is counted and a stopwatch is used to record the duration of the vibrations. The swaying natural frequencies of seven different forms of suspension were measured and are listed in Table 14.2 together with the theoretical predictions obtained using equation 14.10. It can be seen that there is good agreement between the measured and calculated lateral natural frequencies.

Figure 14.4d shows an unstable equilibrium state of the system where the two strings cross each other. An additional horizontal string is applied to prevent out-

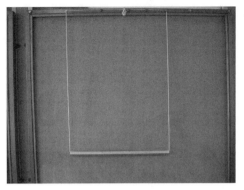

(a) Vertical suspension system

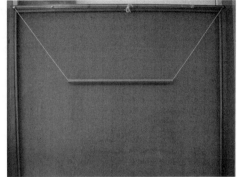

(b) Outward inclined suspension system

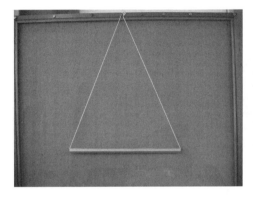

(c) Inward inclined suspension system

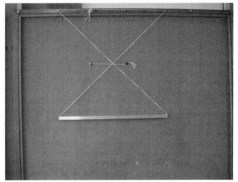

(d) Unstable equilibrium state

Figure 14.4 Typical forms of a suspension system used for demonstration and tests.

Table 14.2 Comparison of the measured and predicted lateral natural frequencies

Parameters	Experiment	Theory	Error (%)
$\theta = 0°$, $l = 0.415$ m	0.778 Hz	0.774 Hz	−0.5
$\theta = -10.15°$, $l = 0.431$ m	0.773 Hz	0.769 Hz	−0.5
$\theta = -22.28°$, $l = 0.825$ m	0.543 Hz	0.536 Hz	−1.3
$\theta = -30.69°$, $l = 0.485$ m	0.718 Hz	0.717 Hz	−0.1
$\theta = -52.7°$, $l = 0.584$ m	UES	UES	−
$\theta = 24.66°$, $l = 0.405$ m	0.930 Hz	0.929 Hz	−0.1
$\theta = 48.4°$, $l = 0.324$ m	1.67 Hz	1.71 Hz	2.4

Note
UES − Unstable equilibrium state.

plane movement of the system. When a small movement is applied to the system, the aluminium bar moves to, and balances at, a new position without oscillation.

14.3.2 Effect of added masses

The models demonstrate that *the natural frequency of a translational suspended system is independent of its mass and the location of the centre of the mass and is the same as that of an equivalent simple pendulum.*

Figure 14.5 shows another simple pendulum system. A plate is suspended by four strings through four holes at the corners of the plate. The other ends of the four strings are fixed to two cantilever frames as shown in Figure 14.5. The strings are vertical when viewed from an angle perpendicular to the plane of the frame and are inclined when viewed in the plane of the frame.

Figure 14.5 A model of a translational suspended system and a rotational suspended system.

When the plate moves in the plane of the steel frames, it forms a translational pendulum system in which the plate remains horizontal during its motion.

Figure 14.6 shows two similar translational suspended systems in which the masses sway in the plane of the supporting frames. Eight magnetic bars and a steel block are placed on the plate of one suspended system to raise the centre of mass of the system. Applying the same displacement to the two plates in the planes of the frames and releasing them simultaneously, it can be observed that the two suspended systems with different masses and different centres of mass sway at the same frequency. This demonstrates that *the natural frequency of a translational suspended system is independent of its mass and the location of the centre of mass.*

Figure 14.6 Effects of mass and the centre of mass.

The suspended systems in Figure 14.6 can also be used as two identical rotational suspended systems when the plates sway perpendicularly to the planes of the frames. Applying the same displacements to the two plates in the direction perpendicular to the frames and then releasing them simultaneously, it can be observed that the plate with the added weights oscillates faster than the other plate.

Figure 14.7 shows two arrangements of eight identical magnetic bars. In one case the eight bars stand on the plate (Figure 14.7a) and in the other case four magnetic bars are placed on the top and bottom surfaces of the plate respectively (Figure 14.7b). The two cases have the same amount of mass but different centres of mass. Conducting the same experiment as before for the rotational suspended systems, it can be seen that the system with standing bars oscillates faster than that with horizontal bars.

The systems shown in Figures 14.5 and 14.7b have different amounts of mass but the same location of the centre of mass. If the oscillations of the two systems shown are generated by giving the same initial displacements in the plane perpendicular to the frames, it can be observed that the two systems oscillate at the same frequency.

The two sets of experiments demonstrate that *the natural frequency of a rotational suspended system is dependent on the location of its centre of mass but is independent of the magnitude of its mass.*

Table 14.3 compares the times recorded for 30 oscillations of the translational and rotational suspended systems with added masses. Each case was tested twice.

(a)   (b)

Figure 14.7 Effect of the centre of mass.

Table 14.3 Comparison of the times for 30 oscillations of the suspended systems

	Translational suspended systems (seconds)	Rotational suspended systems (seconds)
Empty (Figure 14.5)	34.3, 34.4	34.4, 34.5
With full weights (Figure 14.6)	34.3, 34.5	32.3, 32.4
With magnetic bars placed vertically (Figure 14.7a)	34.2, 34.2	33.4, 33.3
With magnetic bars placed horizontally (Figure 14.7b)	34.2, 34.3	34.3, 34.3

14.3.3 Static behaviour of an outward inclined suspended system

This demonstration shows that *an outward inclined suspended system is a mechanism that moves sideways if a vertical load is applied asymmetrically on the plate.*

Consider two suspended systems, where a steel plate is suspended by two vertical strings and the same plate is suspended by two symmetrically outward inclined strings, as shown in Figure 14.8.

Place similar weights asymmetrically on the two plates as shown in Figure 14.8b. It can be observed that the vertically suspended system does not experience any

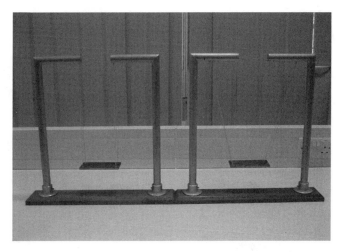

(a)

(b)

Figure 14.8 Static behaviour of an outward inclined suspended system.

lateral movement while the outward inclined suspended system undergoes both lateral and rotational movements.

For the vertically suspended system, the vertical and lateral movements are independent and the vertical load only induces vertical deformations of the strings with little rotation due to the difference of the elastic elongation of the strings. For the plate suspended by the two outward inclined strings, the horizontal and rotational movements of the plate are coupled. The movements relate to the geometry of the system rather than the elastic elongation of the strings.

182 Dynamics

14.4 Practical examples

14.4.1 An inclined suspended wooden bridge in a playground

Figure 14.9 An inclined suspended wooden bridge in a playground.

As demonstrated in section 14.3.3, an inclined suspended system moves sideways if a vertical load is applied asymmetrically. The phenomenon should be avoided in engineering structures, as it may lead to unsafe structures. However, Figure 14.9 shows an outward inclined suspended wooden footbridge in a playground. The bridge is purposely built in such a way so that the bridge wobbles when a child walks on the bridge, creating excitement and a challenge for crossing the bridge.

14.4.2 Seismic isolation of a floor

As the natural frequency of a translational suspended system is only governed by its length and is independent of its mass, the concept of a pendulum seismic isolator was developed and used in Japan.

As shown in Figure 14.10, a floor suspended from girders of a building frame in the form of a translational suspended system was adopted for the exhibition rooms of an actual museum for pottery and porcelain in Japan. The area of the suspended floor is about $1000\,m^2$, and its mass is about 1000 tonnes. Hinges having universal joints were used for the upper and lower ends of the hangers. The hangers were 4.5m long, producing a natural period of more than four seconds, which was considered sufficiently long for seismic isolation.

14.4.3 The Foucault pendulum

Figure 14.11 shows a huge brass pendulum that swings across the lower foyer in the Manchester Conference Centre at the University of Manchester. The pendulum was set up as a tribute to Jean Bernard Leon Foucault.

Pendulums 183

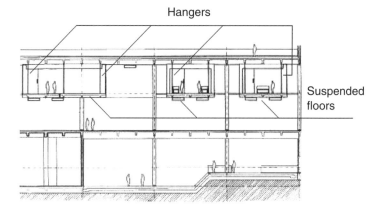

Figure 14.10 A translational suspended floor (courtesy of Professor M. Kawaguchi).

Figure 14.11 The Foucault pendulum.

Jean Bernard Leon Foucault (1819–1868), a French physicist, demonstrated the earth's rotation using his famous pendulum. He suspended a 28 kg bob with a 67 m wire from the dome of the Pantheon in Paris. Foucault was acclaimed by witnesses for proving that the earth does indeed spin on its axis. As the plane of the pendulum oscillation remained unchanged with the stars, they could understand that the Pantheon moved around the pendulum, and not vice versa!

Foucault's pendulum was the first dynamical proof of the earth's rotation in an easy-to-see experiment.

References

14.1 Matthews, M. R., Gauld, C. F. and Stinner, A. (2005) *The Pendulum: Scientific, Historical, Philosophical and Educational Perspectives*, Dordrecht: Springer Netherlands, pp. 37–47.

14.2 Zou, D. and Ji, T. (2007) 'Dynamic characteristics of a generalised suspension system', *International Journal of Mechanical Science*s, Vol. 50, pp. 30–42.

15 Free vibration

15.1 Definitions and concepts

Free vibration: a structure is said to undergo free vibration when it is disturbed from its static stable equilibrium position by an initial displacement and/or initial velocity and then allowed to vibrate without any external dynamic excitation.

Period of vibration: the time required for an undamped system to complete one cycle of free vibration is the natural period of vibration of the system.

Natural frequency: the number of cycles of free vibration of an undamped system in one second is termed the natural frequency of the system, and is the inverse of the period of vibration.

A SDOF system: if the displacement of a system can be uniquely determined by a single variable, this system is called a single-degree-of-freedom (SDOF) system. Normally it consists of a mass, a spring and a damper. The square of the natural frequency of a SDOF system is proportional to the stiffness of the system and the inverse of its mass.

A generalised SDOF system: consider a discrete system that consists of several masses, springs and dampers, or a continuous system that has distributed mass and flexibility. If the shape or pattern of its displacements is known or assumed, the displacements of the system can then uniquely be determined by its magnitude (a single variable). This system is termed a generalised SDOF system. The analysis developed for a SDOF system is applicable to a generalised SDOF system.

- For a structure with a given mass, the stiffer the structure, the higher the natural frequency.
- The larger the damping ratio of a structure, the quicker the decay of its free vibration.
- The higher the natural frequency of a structure, the quicker the decay of its free vibration.
- The fundamental natural frequency reflects the stiffness of a structure. Thus it can be used to predict the displacement of a simple structure. Also the displacement of the structure can be used to estimate its fundamental natural frequency.

15.2 Theoretical background

15.2.1 A single-degree-of-freedom system

Consider a SDOF system, as shown in Figure 15.1a, that consists of:

(a) A SDOF system   (b) Free-body diagram

Figure 15.1 Free vibration of a SDOF system.

- A mass, m, (kg or Ns^2/m), whose motion is to be examined.
- A spring with stiffness k, (N/m). The action of the spring tends to return the mass to its original position of equilibrium. Thus the direction of the elastic force applied on the mass is opposite to that of the motion. Hence the force on the mass is $-ku$ (N).
- A damper (dashpot) that exerts a force whose magnitude is proportional to the velocity of the mass. The constant of proportionality c is known as the *viscous damping coefficient* and has units of Ns/m. The action of the damping force tends to reduce the velocity of motion. Thus the direction of the damping force applied on the mass is opposite to that of the *velocity* of the motion. The force is expressed as $-c\dot{u}$ (N).

Considering the free-body diagram in Figure 15.1b, the equation of motion of the system can be obtained using Newton's second law as:

$$m\ddot{u} = -c\dot{u} - ku \tag{15.1}$$

or

$$m\ddot{u} + c\dot{u} + ku = 0 \tag{15.2}$$

The frequency of oscillation of the system, called its natural frequency in Hz, is:

$$f = \frac{\omega}{2\pi} = \frac{1}{2\pi}\sqrt{\frac{k}{m}} \tag{15.3a}$$

For viscous damping, it can be shown that:

$$c = 2\xi m\omega \tag{15.3b}$$

where ξ is the damping ratio. Substituting equation 15.3 into equation 15.2 gives:

$$\ddot{u} + 2\xi\omega\dot{u} + \omega^2 u = 0 \tag{15.4}$$

The solution of equation 15.4 is in the form of [15.1, 15.2]:

$$u = Ae^{s_1 t} + Be^{s_2 t} \qquad (15.5a)$$

$$s_{1,2} = [-\xi \pm \sqrt{\xi^2 - 1}]\omega \qquad (15.5b)$$

where A and B are constants that can be determined from the initial conditions. It can be seen from equation 15.5 that the response of the system depends on whether the damping ratio ξ is greater than, equal to or smaller than unity.

Case 1: $\xi = 1$, i.e. critically-damped systems.
In this special case, there are two identical roots from equation 15.5b:

$$s_1 = s_2 = -\omega \qquad (15.6)$$

Substituting equation 15.6 into equation 15.5a gives:

$$u = (A + Bt)e^{-\omega t} \qquad (15.7)$$

If the mass is displaced from its position of static equilibrium at time $t = 0$, the initial conditions $u(0)$ and $\dot{u}(0)$ can be used to determine the two constants in equation 15.7. Thus equation 15.7 can be written as:

$$u = [u(0)(1 + \omega t) + \dot{u}(0)t]e^{-\omega t} \qquad (15.8)$$

Equation 15.8 can be illustrated graphically in Figure 15.2a for the values of $u(0) = 1$ cm, $\dot{u}(0) = 1$ cm/s and $\omega = 1$ rad/s. This shows that the response of the critically-damped system does not oscillate about its equilibrium position, but returns to the position of equilibrium asymptotically, as dictated by the exponential term in equation 15.8.

It is of interest to know the condition if the response crosses the zero-deflection position. Solving equation 15.8 at $u = 0$ gives the condition that the response crosses the horizontal axis once at time:

$$t = -\frac{u(0)}{\dot{u}(0) + u(0)\omega} \qquad (15.9)$$

Consider a case where $u(0) = 1$ cm, $\dot{u}(0) = -1.4$ cm/s and $\omega = 1$ rad/s and where t must be larger than zero. Substituting the values into equation 15.9 gives the solution $t = 2.5$ s. Figure 15.2b shows the curve defined by equation 15.8 for this particular case. It can be seen that the response crosses the zero-deflection position at 2.5 s then monotonically approaches the zero-deflection position.

Case 2: $\xi > 1$, i.e. overcritically-damped systems.
In this special case, the solution given in equation 15.5 can be re-written as [15.2, 15.3]:

$$u(t) = [A\sinh\bar{\omega}t + B\cosh\bar{\omega}t]e^{-\xi\omega t} \qquad (15.10)$$

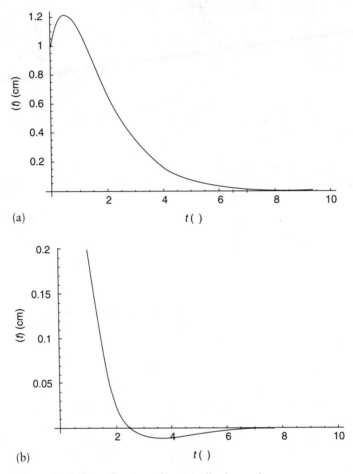

Figure 15.2 Free vibration of a critically-damped system.

where $\bar{\omega} = \omega\sqrt{\xi^2 - 1}$. The constants A and B can be determined using the initial conditions $u(0)$ and $\dot{u}(0)$ and equation 15.10 becomes:

$$u(t) = \left[\frac{\dot{u}(0) + u(0)\xi\omega}{\bar{\omega}} \sinh\bar{\omega}t + u(0)\cosh\bar{\omega}t \right] e^{-\xi\omega t} \tag{15.11}$$

Equation 15.11 shows that the response reduces exponentially, which is similar to the motion of the critically-damped system. For example, using the same data $u(0) = 1\,\text{cm}$, $\dot{u}(0) = 1\,\text{cm/s}$ and $\omega = 1\,\text{rad/s}$ as were used for producing Figure 15.2a, but with $\xi = 2$. The free vibration of the overcritically-damped system is shown in Figure 15.3.

Comparing the displacement-time histories in Figures 15.2a and 15.3, it can be noted that the overcritically-damped system takes a longer time to return to the original equilibrium position than the critically-damped system. The movement of an overcritically-damped system will be demonstrated in section 15.3.3.

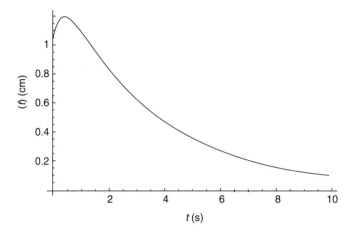

Figure 15.3 Free vibration of an overcritically-damped system.

Case 3: $\xi < 1$, i.e. undercritically-damped systems.
In this special case, the two roots in equation 15.5b become:

$$s_{1,2} = -\xi\omega \pm i\omega_D \tag{15.12}$$

where

$$\omega_D = \omega\sqrt{1-\xi^2} \tag{15.13}$$

ω_D is the damped natural frequency of the system in free vibrations. The solution given in equation 15.5 becomes:

$$u(t) = [A\cos\omega_D t + B\sin\omega_D t]e^{-\xi\omega t} \tag{15.14}$$

Using the initial conditions $u(0)$ and $\dot{u}(0)$, the constants can be determined leading to:

$$u(t) = \left[\frac{\dot{u}(0) + u(0)\xi\omega}{\omega_D}\sin\omega_D t + u(0)\cos\omega_D t\right]e^{-\xi\omega t} \tag{15.15}$$

or

$$u(t) = d\cos(\omega_D t - \theta)e^{-\xi\omega t} \tag{15.16}$$

where

$$d = \sqrt{u(0)^2 + \left(\frac{\dot{u}(0) + u(0)\xi\omega}{\omega_D}\right)^2} \tag{15.17a}$$

$$\sin\theta = \frac{\dot{u}(0) + u(0)\xi\omega}{\omega_D d}; \quad \cos\theta = \frac{u(0)}{d} \tag{15.17b, c}$$

190 Dynamics

$$\theta = \tan^{-1}\left\{\frac{\dot{u}(0)+u(0)\xi\omega}{\omega_D u(0)}\right\} \qquad (15.18)$$

Figure 15.4 shows a typical free vibration of an undercritically-damped SDOF system defined by equation 15.16.

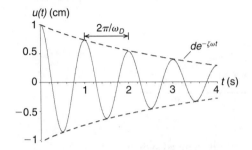

Figure 15.4 Free vibration of an undercritically-damped system.

From Figure 15.4 and equation 15.15 it can be seen that:

- The damped system oscillates about the position of equilibrium with a damped natural frequency of ω_D.
- The oscillation decays exponentially.
- The rate of the exponential decay depends on the product of the damping ratio ξ and the natural frequency of ω.

This last point indicates that *a system with a higher natural frequency will decay more quickly than a similar system with a lower natural frequency in free vibration.*

Example 15.1

Two lightly damped SDOF systems have the same damping ratio $\xi = 0.05$ (or 5 per cent critical) but different natural frequencies of $\omega_1 = 2\pi$ rad/s and $\omega_2 = 4\pi$ rad/s respectively. Applying the same initial displacement $u(0) = 1.0$ cm to the two systems and releasing them simultaneously will generate free vibrations. Calculate the exponential decay at $t = 3$ s and plot the vibration time histories.

Solution

For the first system: $\quad -\xi\omega_1 t = -0.05 \times 2\pi \times 3 = -0.3\pi = -0.942$
The exponential decay is: $e^{-\xi\omega_1 t} = e^{-0.942} = 0.390$
For the second system: $\quad -\xi\omega_2 t = -0.05 \times 4\pi \times 3 = -0.6\pi = -1.885$
The exponential decay is: $e^{-\xi\omega_2 t} = e^{-1.885} = 0.152$

Figure 15.5 shows the curves of free vibrations of the two systems using equation 15.15. The results show that *the higher the natural frequency, the quicker the decay of its free vibration.*

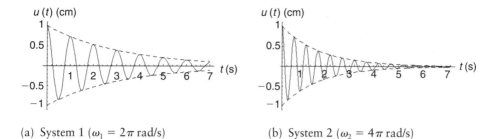

(a) System 1 ($\omega_1 = 2\pi$ rad/s)   (b) System 2 ($\omega_2 = 4\pi$ rad/s)

Figure 15.5 Comparison of the decay of free vibration of two systems with different natural frequencies.

15.2.2 A generalised single-degree-of-freedom system

Consider the vibration of a particular mode of a structure, the vibration can be described by a SDOF system. Thus it is necessary to calculate the generalised properties of the system, i.e. the generalised mass, damping and stiffness.

Consider a continuous system, such as the simply supported beam shown in Figure 15.6a. For a particular mode of vibration $\phi(x)$, which may be known or, if not, be assumed with a maximum value of unity, the vibration of the mode can be described by the variable $z(t)$ at the centre of the beam and the movement of the beam at coordinate x as $z(t)\phi(x)$. The vibration of the beam $v(x, t)$ in the particular mode can then be represented by:

$$v(x, t) = z(t)\phi(x) \tag{15.19}$$

The generalised properties for the vibration of the system in the particular mode are [15.2]:

$$\text{Modal mass: } m^* = \int_0^L m(x)\phi(x)^2 dx \tag{15.20a}$$

$$\text{Modal stiffness: } k^* = \int_0^L EI(x)\phi''(x)^2 dx \tag{15.20b}$$

$$\text{Damping coefficient: } c^* = 2\xi m^* \omega \tag{15.21}$$

The equation of motion of the system (equation 15.2) then becomes:

$$m^*\ddot{z} + c^*\dot{z} + k^*z = 0 \tag{15.22}$$

The natural frequency of the vibration mode of the structure can be obtained from equation 15.22 as follows:

$$\omega^2 = \frac{k^*}{m^*} = \frac{\int_0^L EI(x)\phi''(x)^2 dx}{\int_0^L m(x)\phi(x)^2 dx} \tag{15.23}$$

192 Dynamics

The accuracy of the natural frequency of the vibration mode depends on the quality of the known or assumed mode shape $\phi(x)$. In order to obtain a good estimation of the natural frequency, the assumed shape $\phi(x)$ should satisfy as many boundary conditions of the structure as possible. In general:

1. For a pinned support at $x=0$, the displacement and bending moment at the support should be zero, i.e.:

   $\phi(0) = 0$ and $\phi''(0) = 0$

2. For a fixed support at $x=0$, the displacement and rotation at the support should be zero, i.e.:

   $\phi(0) = 0$ and $\phi'(0) = 0$

3. For a free end at $x=0$, the bending moment and shear force at the free end should be equal to the applied load P and bending moment M, i.e.:

   $EI\phi''(0) = -M$ and $EI\phi'''(0) = -P$

   If there is no applied loading at the free end, $M = P = 0$.

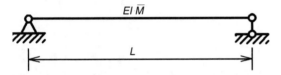

(a) A simply supported beam

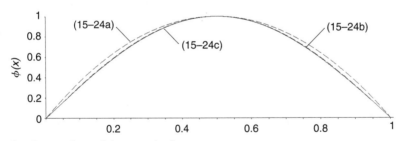

(b) Comparison of three mode shapes

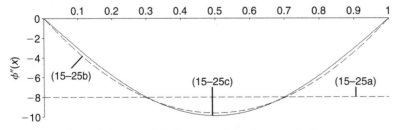

(c) Comparison of the second derivatives of the three mode shapes

Figure 15.6 A simply supported beam with three fundamental mode shapes considered.

Free vibration 193

A good approximation of the fundamental mode shape is the shape of static deflection of a structure when it is subjected to distributed loads proportional to the mass distribution of the structure.

Example 15.2

Figure 15.6a shows a simply supported beam with a length of L, uniformly distributed mass of $\overline{m}$ and flexural rigidity of EI. Three different mode shapes will be considered in this analysis. They are:

1    $\phi_1(x) = 4(x/L)(1 - x/L)$ (15.24a)

2    $\phi_2(x) = \dfrac{16x}{5L^4}(L^3 - 2Lx^2 + x^3)$ (15.24b)

3    $\phi_3(x) = \sin\dfrac{\pi x}{L}$ (15.24c)

Calculate and compare the modal masses, modal stiffnesses and natural frequencies determined using the three mode shapes.

Solution

Figure 15.6b compares the three mode shapes used in equation 15.24 where the horizontal axis is x/L, varying between 0 and 1. It can be seen that the curves defined by equation 15.24b and equation 15.24c almost overlap although there are small differences between the curves defined by equation 15.24a and equations 15.24b and 15.24c.

Differentiating equations 15.24a, 15.24b and 15.24c twice with respect to x gives:

$$\phi_1''(x) = -\frac{8}{L^2} \qquad (15.25a)$$

$$\phi_2''(x) = -\frac{192x(L-x)}{5L^4} \qquad (15.25b)$$

$$\phi_3''(x) = -\frac{\pi^2}{L^2}\sin\frac{\pi x}{L} \qquad (15.25c)$$

Figure 15.6c compares the shapes of the three second derivatives. The curves defined by equation 15.25b and equation 15.25c are close to each other. However, there are significant differences between the curves defined by equation 15.25a and equations 15.25b and 15.25c.

Case 1: this assumed shape satisfies $\phi(0) = 0$ and $\phi(L) = 0$. However, $\phi''(0) \neq 0$ and $\phi''(L) \neq 0$. Using equations 15.20 and 15.23 gives:

$$m^* = \overline{m} \int_0^L \left[\frac{4x}{L}\left(1-\frac{x}{L}\right)\right]^2 dx = \frac{8\overline{m}L}{15}$$

$$k^* = EI \int_0^L \phi''(x)^2 = EI \int_0^L \left(-\frac{8}{L^2}\right)^2 dx = \frac{64EI}{L^3}$$

$$\omega^2 = \frac{k^*}{m^*} = \frac{64EI}{L^3} \frac{15}{8\overline{m}L} = \frac{120EI}{\overline{m}L^4} \qquad (15.26a)$$

Case 2: equation 15.24b is the shape of the static deflection curve when the beam is subjected to a uniformly distributed load. This function satisfies the required boundary conditions and:

$$m^* = \overline{m} \int_0^L \phi(x)^2 dx = \frac{3968\overline{m}L}{7875}$$

$$k^* = EI \int_0^L \phi''(x)^2 = EI \int_0^L \left(-\frac{192x(L-x)}{5L^4}\right)^2 dx = \frac{6144EI}{125L^3}$$

$$\omega^2 = \frac{k^*}{m^*} = \frac{6144EI}{125L^3} \frac{7875}{3968\overline{m}L} = \frac{3024EI}{31\overline{m}L^4} = \frac{97.55EI}{\overline{m}L^4} \qquad (15.26b)$$

Case 3: equation 15.24c is the true shape of the fundamental mode of a simply supported beam and leads to:

$$m^* = \overline{m} \int_0^L \phi(x)^2 dx = \frac{\overline{m}L}{2}$$

$$k^* = EI \int_0^L \phi''(x)^2 = EI \int_0^L \left(-\frac{\pi^2}{L^2} \sin\frac{\pi x}{L}\right)^2 dx = \frac{\pi^4 EI}{2L^3}$$

$$\omega^2 = \frac{k^*}{m^*} = \frac{\pi^4 EI}{2L^3} \frac{2}{\overline{m}L} = \frac{\pi^4 EI}{\overline{m}L^4} = \frac{97.41EI}{\overline{m}L^4} \qquad (15.26c)$$

Table 15.1 Comparison of coefficients of m^*, k^* and ω^2

	$k^* = a_1 EI/L^3$		$m^* = a_2 \overline{m}L$		$\omega^2 = a_3 EI/(\overline{m}L^4)$	
	a_1	R.E.	a_2	R.E.	a_3	R.E.
Equation 15.26a	64	31.4%	$\frac{8}{15}$	6.7%	120	23.2%
Equation 15.26b	$\frac{6144}{125}$	0.9%	$\frac{3968}{7875}$	0.8%	$\frac{3024}{31}$	0.1%
Equation 15.26c	$\frac{\pi^4}{2}$	0%	$\frac{1}{2}$	0%	π^4	0%

Free vibration

Table 15.1 compares coefficients of m^*, k^* and ω^2 in equation 15.26 and gives the relative errors (R.E.) against the exact values. It can be seen from Table 15.1 that:

- Equation 15.26b produces a solution very close to the true solution given by equation 15.26c.
- The solution obtained using equation 15.26a is 23 per cent larger than the true solution, which is because the assumed mode shape does not fully satisfy the boundary conditions.
- The differences between the modal stiffnesses are larger than the differences between the modal masses, as there are small differences between the three shape functions which are used for calculating the modal masses, but there are larger differences between the second derivatives of the three functions which are used to calculate the modal stiffnesses.
- Using the assumed mode shapes overestimates the modal stiffness, modal mass and natural frequency of the beam.

Conceptually, the use of assumed mode shapes rather than the true mode shapes is equivalent to applying additional external constraints on the structure to force the structure to deform in the assumed mode shape. These constraints effectively stiffen the structure, leading to an increased stiffness, and hence overestimate the true natural frequencies. This indicates that the lowest natural frequency obtained using a number of different assumed mode shapes would be the closest to the true value. For instance, the estimated natural frequency in case 2 is smaller than that in case 1, thus case 2 provides a better estimation than case 1.

15.2.3 A multi-degrees-of-freedom (MDOF) system

Consider the free vibration of a linear system that consists of N degrees-of-freedom. The equation of undamped free vibration of the system has the following form:

$$M\ddot{v} + Kv = 0 \tag{15.27}$$

where M is the mass matrix, K is the stiffness matrix and v is the vector of displacements. The natural frequencies and mode shapes of the system can be obtained by solving the eigenvalue problem:

$$[K - \omega_i^2 M]\phi_i = 0 \tag{15.28}$$

where ω_i and ϕ_i are the natural frequency and shape of the ith mode ($i = 1, 2, \ldots, N$). ω_i^2 is also known as an eigenvalue or characteristic value.

The N modes obtained from equation 15.28 are independent, i.e. no one mode can be expressed by a linear combination of the other modes. These modes satisfy the following orthogonality conditions:

$$\phi_i^T K \phi_j = 0 \quad \phi_i^T M \phi_j = 0 \quad i \neq j \tag{15.29a}$$

$$\phi_i^T K \phi_i = K_i \quad \phi_i^T M \phi_i = M_i \quad i = j \tag{15.29b}$$

where K_i and M_i are the modal stiffness and modal mass of the ith mode and are positive, and related by:

$$K_i = \omega_i^2 M_i \qquad (15.30)$$

The summation of the response of several modes of a structure can be used to represent the structural response. As the N independent mode vectors form a base in the N dimensional space, any other vector in the space can be expressed as a linear combination of the N mode vectors, i.e.:

$$v = z_1\phi_1 + z_2\phi_2 + \ldots + z_N\phi_N = \sum_{i=1}^{N} z_i\phi_i \qquad (15.31)$$

where z_i is the magnitude of the ith mode, defining the contribution of the mode to the total response. In practice, it is frequently the case that only the first few modes can reasonably represent the total dynamic response of structures. This leads to an effective simplification of analysis without significant loss of accuracy. Figure 15.7 shows the vibration of a cantilever column. Its deflected shape can be reasonably represented by the summation of the response of the first three modes. (The shapes of the first two modes of vibration of a cantilever will be shown in section 15.3.5.)

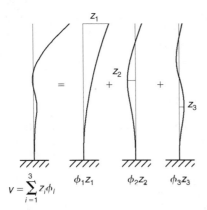

Figure 15.7 Deflected shape as the sum of modal components [15.2].

15.2.4 Relationship between the fundamental natural frequency and the maximum displacement of a beam

The fundamental natural frequency of a uniform beam can be expressed using equations 15.23 and 15.26:

$$f = \frac{\omega}{2\pi} = \frac{b_1}{2\pi}\sqrt{\frac{EI}{mL^4}} \qquad (15.32)$$

where b_1 is a constant related to the boundary conditions. The values of b_1 for beams with common boundary conditions are given in Table 15.2. For example,

$b_1 = \pi^2$ for a simply supported beam. From Chapter 7 it is known that the maximum displacement of the uniform beam, listed in Table 15.2, subjected to self-weight can be expressed in a unified form as follows:

$$\Delta = d_1 \frac{\overline{m}gL^4}{EI} \qquad (15.33)$$

where g is gravity, $\overline{m}$ is the distributed mass of the beam along its length L and d_1 is a constant dependent on the boundary conditions. For example, $d_1 = 5/384$ for a simply supported beam. The constant d_1 for the beams with four common boundary conditions are given in Table 15.2.

Table 15.2 Coefficients for single-span beams

	Simply supported beam	Fixed end beam	Propped cantilever beam	Cantilever beam
b_1	9.87	22.4	15.4	3.52
d_1	0.0130	0.00260	0.00542	0.125
$A_1(m^{1/2}/s)$	0.561	0.570	0.565	0.621
$B_1(m/s^2)$	0.315	0.325	0.320	0.385

It can be noted that equations 15.32 and 15.33 contain the same term EI, through which a relationship between the fundamental natural frequency f and the maximum displacement Δ due to the self-weight of a beam can be established. Equations 15.32 and 15.33 can be expressed as:

$$EI = \frac{4\pi^2 \overline{m} L^4}{b_1^2} f^2 \qquad EI = d_1 \frac{\overline{m}gL^4}{\Delta}$$

leading to:

$$f = \frac{b_1}{2\pi} \sqrt{d_1 g} \sqrt{\frac{1}{\Delta}} = A_1 \sqrt{\frac{1}{\Delta}} \qquad (15.34)$$

$$\Delta = \frac{b_1^2 d_1 g}{4\pi^2} \frac{1}{f^2} = B_1 \frac{1}{f^2} \qquad (15.35)$$

As the constants b_1 and d_1 are given, the coefficients A_1 and B_1 can be calculated for the beams listed in Table 15.2 and have units of $m^{1/2}/s$ and m/s^2 respectively. The corresponding values of A_1 and B_1 are given in Table 15.2 using $g = 9.81 m/s^2$. Thus the displacement Δ is measured by m in the above two equations. The fundamental natural frequency of a beam can be calculated using Table 15.2 and equation 15.34 if the maximum displacement of the beam due to its self-weight is known. Alternatively, the maximum displacement of a beam can be estimated using Table 15.2 and equation 15.35 if the fundamental natural frequency is available either from calculation or vibration measurement. Examples 15.3 and 15.4 illustrate the applications.

198 Dynamics

Example 15.3

A cantilever beam supporting a floor has a self-weight of 20 kN/m. A static analysis of the beam shows the maximum displacement of 0.06 m subject to live load of 30 kN/m. Calculate the fundamental natural frequency of the beam.

Solution

The maximum displacement of the structure subjected to its self-weight should be $0.06 \times (20/30) = 0.04$ m. For a cantilever beam, select $A_1 = 0.621$ from Table 15.2. The fundamental natural frequency of the cantilever beam is thus calculated as:

$$f = A_1 \sqrt{\frac{1}{\Delta}} = 0.621 \sqrt{\frac{1}{0.04}} = 3.11 \, \text{Hz} \tag{15.36}$$

Example 15.4

A simply supported beam bridge has a self-weight of 30 kN/m and the fundamental frequency of the bridge of 4.5 Hz. Estimate the maximum displacement of the bridge for the live load of 150 kN/m.

Solution

For a simply supported beam, select $B_1 = 0.315$ from Table 15.2. Using equation 15.35 gives the maximum displacement of the bridge for self-weight only as:

$$\Delta = B_1 \frac{1}{f^2} = 0.315 \frac{1}{4.5^2} = 0.0156 \, \text{m}$$

The maximum displacement due to the live load will be:

$0.0156 \times 150/30 = 0.078$ m

It is interesting to note in Table 15.2 that the values of A_1 and B_1 for the four cases are close, in particular for the first three cases; indicating that A_1 and B_1 are not sensitive to boundary conditions. This observation is useful for practical application of equations 15.34 and 15.35 as boundary conditions are often not either truly pinned or fixed. Actually, the effect of boundary conditions have been considered in Δ for determining the fundamental natural frequency f, and in f for calculating the maximum displacement Δ.

15.2.5 Relationship between the fundamental natural frequency and the tension force in a straight string

The vibration of a straight string in tension can be described by the following equation [15.4]:

$$\overline{m} \frac{\partial^2 v}{\partial t^2} = F \frac{\partial^2 v}{\partial x^2} \tag{15.37}$$

Free vibration

where v is the transverse vibration of the string, and is a function of time t and position x. F is the tension force in the string and $\overline{m}$ is the distributed mass of the string.

Considering the vibration of the first mode:

$$v(x, t) = A\sin\frac{\pi x}{L}\sin(2\pi f)t \tag{15.38}$$

where L is the length of the string and f is the fundamental natural frequency of the string in the transverse direction. Substituting equation 15.37 into equation 15.36 gives:

$$F = 4\overline{m}L^2 f^2 \tag{15.39}$$

Equation 15.38 indicates that:

- The tension force in the string can be calculated using the fundamental natural frequency obtained from either measurement or analysis.
- The tension force in the string is proportional to the total mass of the string $\overline{m}L$, its span L and its natural frequency squared, f^2.

Similar formulae to equation 15.38 are available for predicting the tension force in a string or a cable [15.5]. Section 15.3.6 will provide a model test and section 15.4.6 will show a practical application.

15.3 Model demonstrations

15.3.1 Free vibration of a pendulum system

This demonstration shows *the definitions of natural frequency and period of a simple system.*

Figure 15.8 shows a pendulum system consisting of five balls suspended by strings with equal length (known as Newton's cradle). Push the balls in a direction parallel to the plane of the frames to apply an initial displacement and release the balls. The balls then move from side to side in free vibration at the natural frequency of the system.

Count the number of oscillations of the balls in 30 seconds and the number of oscillations divided by 30 is the natural frequency of the oscillations of the pendulum system in cycles per second or Hz. The inverse of the natural frequency is the period of the oscillations. For the studied system, 40 cycles of oscillation in 30 seconds were counted. Thus the natural frequency of the system is $40/30 = 1.33\,\text{Hz}$ and the period is $1/1.33 = 0.75\,\text{s}$.

The vertical distance between the supports of the strings and the connecting points on the balls is 140 mm so the theoretical value of the natural frequency is:

$$f = \frac{1}{2\pi}\sqrt{\frac{g}{L}} = \frac{1}{2\pi}\sqrt{\frac{9810}{140}} = 1.33\,\text{Hz} \tag{14.3}$$

200 *Dynamics*

Figure 15.8 Free vibration and natural frequency of a pendulum system.

15.3.2 Vibration decay and natural frequency

This set of models demonstrates that *the higher the natural frequency of a structure, the quicker the decay of its free vibration.*

A single steel ruler and a pair of identical steel rulers, bolted firmly together, are placed side by side as cantilevers shown as in Figure 15.9a. Give the ends of the two cantilevers the same initial displacement and release them suddenly at the same time (Figure 15.9a). Free vibrations of the rulers follow and it will be observed that the bolted double ruler stops vibrating much more rapidly than the single ruler (Figure 15.9b).

(a) Applying the same initial displacement

(b) The double ruler stops vibration more quickly than the single ruler

Figure 15.9 The decay of free vibration and the natural frequencies of members.

The rate of decay of free vibration is proportional to the product of the damping ratio and the natural frequency of the structure as shown in equation 15.15 and example 15.1. The bolted double ruler has a higher natural frequency since its second moment of area is eight times of that of the single ruler, making it eight times as stiff as the single ruler, while its mass is just double of that of the single ruler. In fact the fundamental natural frequency of the bolted double ruler is twice that of the single ruler. The damping ratios for the two cantilevers can be considered to be the same. Hence this demonstration verifies the concept that *the higher the natural frequency of a structure, the quicker the decay of its free vibration.*

15.3.3 An overcritically-damped system

This demonstration shows *the movement rather than the vibrations of an overcritically-damped beam subject to an initial displacement.*

Figure 15.10 shows two cantilever metal strips representing two cantilever beams. The one on the left is a conventional metal strip and the one on the right uses two metal strips with a layer of damping material constrained between the two strips. Apply the same initial displacement on the free ends of the two cantilevers and release them at the same time. It is observed that the cantilever on the right returns to its original position slowly without experiencing any vibration. This is the phenomenon of an overcritically-damped system described in section 15.2.1 and Figure 15.3.

Figure 15.10 Demonstration of an overcritically-damped system.

202 Dynamics

15.3.4 Mode shapes of a discrete model

This demonstration shows *the two mode shapes of a discrete TDOF system.*

Figure 15.11 shows a two-degree-of-freedom (TDOF) model that has two natural frequencies and two vibration modes. Apply initial horizontal displacements of the two masses in the same direction and then release them at the same time, the first mode of vibration is generated, Figure 15.11a. Apply initial horizontal displacements of the two masses in opposite directions and then release them at the same time, the second mode of vibration is generated, Figure 15.11b.

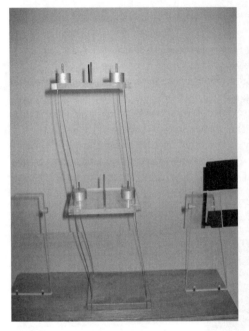 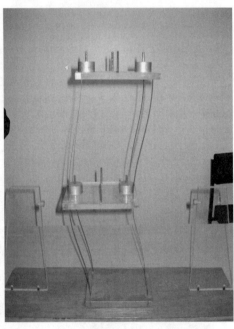

(a) The first mode of vibration   (b) The second mode of vibration

Figure 15.11 Mode shapes of a TDOF model.

15.3.5 Mode shapes of a continuous model

This demonstration shows *the shapes of the first two modes of vibration of a continuous beam.*

Take a long plastic strip and hold one end of the strip as shown in Figure 15.12. Then moving the hand forward and backward slowly, it generates the first mode of vibration as shown in Figure 15.12a. Do the same but increase the speed of the movement, it excites the second mode of vibration as given in Figure 15.12b. The two mode shapes of the strip are similar to those shown in Figure 15.7.

Free vibration 203

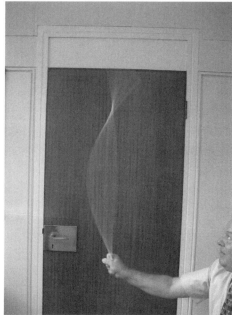

(a) The first mode of vibration (b) The second mode of vibration

Figure 15.12 Mode shapes of a cantilever.

15.3.6 *Tension force and natural frequency of a straight tension bar*

This experiment *verifies equation 15.38* and shows that *the force in a tension bar can be predicted using natural frequency measurement*.

Two steel bars with different diameters (6 mm and 8 mm) and lengths (2.55 m and 2.73 m) were fixed to the steel frame shown in Figure 15.13a. Two strain gauges

(a) (b)

Figure 15.13 Verification test of a tension bar.

204 Dynamics

were glued on each of the bars opposite to each other at the mid-points of the bars. Tension forces were applied through turnbuckles at the ends of the bars, and the values of forces were determined through strain measurements. A small accelerometer was fixed to one of the two bars as shown in Figure 15.13b to record vibrations.

Vibration tests of the tension bars were conducted and accelerations were recorded when the bars were tensioned to different levels. Fundamental natural frequencies were determined from spectral analysis of the acceleration responses. For a range of measured fundamental natural frequencies Table 15.3 compares the measured tension forces in the 6mm tension bar and the forces predicted using equation 15.38 where the measured fundamental natural frequency values were taken. For predictions, the following values were used: $L = 2.55\,\text{m}$, $\rho = 7850\,\text{kg/m}^3$ and $A = \pi r^2 = \pi (0.003)^2\,\text{m}^2$.

Table 15.3 Comparison of the measured and predicted tension forces for 6mm bar [15.5]

Experimental results		Prediction using equation 15.38	
Natural frequency (Hz)	Tension force (kN)	Tension force (kN)	Relative error (%)
13.9	1.16	1.16	0.00
15.5	1.42	1.39	2.11
16.9	1.70	1.65	2.94
18.3	1.98	1.93	2.53
19.6	2.25	2.22	1.33
20.5	2.50	2.43	2.80
21.9	2.77	2.77	0.00
22.8	3.05	3.00	1.67

It can be seen from Table 15.3 that the tension forces in a tensioned bar can be predicted using equation 15.38 and the natural frequency measurements.

15.4 Practical examples

15.4.1 A musical box

A musical box is a device that produces music using mechanical vibration. Figure 15.14a shows one of many decorative music boxes which are readily available. The core of the music box is the unit shown in Figure 15.14b.

A spring is used to rotate a music tube, converting the potential energy stored in the spring into the kinetic energy which drives the rotation of the tube. Raised points on the tube displace cantilever metal bars causing them to vibrate and generate sound. Different geometries (lengths and cross-sections) of the bars (see Figure 15.15) provide different natural frequencies of the bars, generating different music notes. The distribution of the raised points on the tube is designed to create a particular music tune when the tube rotates. The music unit shown in Figure 15.14b has 18 metal bars generating 18 different musical notes.

(a) A musical box (b) The core of the musical box

Figure 15.14 A musical box and its core (the models are provided by Professor B. Zhuang, Zhejiang University, China).

 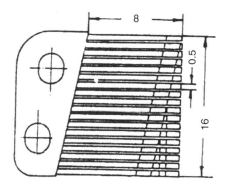

Figure 15.15 Cantilever beams with different lengths and sections in the music box (mm) [15.6].

A given musical note relates to a particular natural frequency of the vibrating body. Table 15.4 gives the relationships between some natural frequencies and music notes.

Figure 15.16 shows part of the keyboard of a piano. The main reference point on a piano is known as Middle C. This is the white note located approximately in the centre of the keyboard and immediately to the left of a pair of black keys. Striking the Middle C key, the sound generated corresponds to a frequency of 262 Hz. The next white key, D, to the right of the Middle C key, produces a sound corresponding to a frequency of 294 Hz.

Table 15.4 Relations between natural frequency and music note [15.6, 15.7]

Natural frequency (Hz)	261.6	293.7	329.6	349.2	392	440	493.9
Music note	(Middle) C	D	E	F	G	A	B

Figure 15.16 The keyboard of a piano.

15.4.2 Measurement of the fundamental natural frequency of a building through free vibration generated using vibrators

When the free vibrations of a structure can be measured as accelerations, velocities or displacements, the natural frequencies of the structure can be determined from their vibration time histories.

Figure 15.17 shows an eight-storey steel-framed test building in the Cardington Laboratory of the Building Research Establishment Ltd. Different dynamic test

Figure 15.17 The test building (courtesy of Building Research Establishment Ltd).

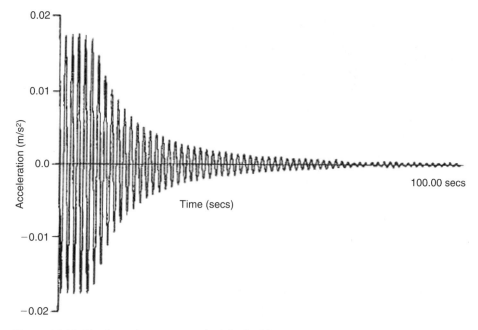

Figure 15.18 The free vibration record of the building.

methods were used to determine the natural frequencies of this building. One of the methods used was to record the free vibrations of the building [15.8].

Four vibrators mounted at the corners of the roof of the building were used to generate movement of the structure. Once movement was initiated the vibrators were turned off and the resulting free vibrations of the structure were measured by accelerometers. Figure 15.18 shows the acceleration–time history of the decayed vibrations. The frequency of the oscillations, which is a natural frequency of the structure, was determined from the time history as the inverse of the time interval between two successive acceleration peaks.

15.4.3 Measurement of the natural frequencies of a stack through vibration generated by the environment

Environmental effects such as air movements around a structure can also induce vibrations, though such vibration may be very small. These types of vibration may not be exactly free vibration as they are caused by disturbances due to external effects, such as wind. However, the concept of free vibration can still be used to identify the natural frequencies of a structure.

Figure 15.19 shows a 97.5 m tall multiflare stack which can be used to burn off excess gases. A laser test system was set up approximately 100 m from the stack to monitor the stack vibrations. Several velocity–time histories on selected measurement points of the stack were measured when wind was blowing.

The natural frequencies of 0.67 Hz and 0.73 Hz were measured which corresponded to the fundamental modes in the two orthogonal horizontal directions.

Figure 15.19 A multiflare stack.

15.4.4 *The tension forces in the cables in the London Eye*

The wheel of the London Eye is stiffened by cables as shown in Figure 15.20. There are 16 rim rotation cables, each around 60 mm thick. In addition, there are 64 spoke cables, which are all 70 mm thick and are spun from 121 individual strands in layers. Part of the tension in the cables will be lost due to normal operations after a period of time. Thus the cables need to be re-tensioned to maintain their design values. It is not convenient to measure the tension force directly in each of the cables. Instead, transverse impact loads are applied to each cable to generate free vibrations, from which the fundamental natural frequency of the cable in the transverse direction can be determined. Using the measured fundamental natural frequency, the tension in each cable can be predicted using equation 15.38 or similar equations. Tension losses can then be identified and recovered.

Free vibration 209

Figure 15.20 The London Eye.

References

15.1 Beards, C. F. (1996) *Structural Vibration: Analysis and Damping*, London: Arnold.
15.2 Clough, R. W. and Penzien, J. (1993) *Dynamics of Structures*, New York: McGraw-Hill.
15.3 Chopra, A. K. (1995) *Dynamics of Structures*, New Jersey: Prentice Hall Inc.
15.4 Morse, P. M. (1948) *Vibration and Sound*, New York: McGraw-Hill.
15.5 Tzima, K. (2003) 'Predicting tension of a stayed cable using frequency measurements', MSc Dissertation, UMIST.
15.6 Zhuang, B., Zhu, Y., Bai, C., Cui, J. and Zhuang, B. (1996) *Design of Music Boxes and Vibration of Tuning Bars*, China: New Times Press.
15.7 White, H. E. and White, D. H. (1980) *Physics and Music: The Science of Musical Sound*, Philadelphia: Saunders College.
15.8 Ellis, B. R. and Ji, T. (1996) 'Dynamic testing and numerical modelling of the Cardington steel framed building from construction to completion', *The Structural Engineer*, Vol. 74, No. 11, pp. 186–192.

16 Resonance

16.1 Definitions and concepts

Resonance is a phenomenon which occurs when the vibration of a system tends to reach its maximum magnitude. The frequency corresponding to resonance is known as the resonance frequency of the system. When the damping of the system is small, the resonance frequency is approximately equal to the natural frequency of the system.

- The resonance frequency is related to the damping ratio, the input (loading or ground motion) and the selected measurement parameter (relative or absolute movement, displacement, velocity or acceleration).
- Increasing damping will effectively reduce the response of the structure at resonance.

For a single-degree-of-freedom (SDOF) system subject to a harmonic input, such as a direct load or a base motion, there are three characteristics:

- The maximum dynamic displacement is close to the static displacement for a given load amplitude if the load frequency is less than a quarter of the natural frequency of the system.
- The maximum dynamic displacement is less than the static displacement if the load frequency is more than twice the natural frequency of the system.
- The maximum dynamic displacement is several times the static displacement if the load frequency is close to or matches the natural frequency of the system.

16.2 Theoretical background

Although the above statements are abstracted from the study of a single-degree-of-freedom system subject to a harmonic input, they are applicable to many practical situations. This is because:

- The response of a structure can be expressed as the summation of the responses of several modes of vibration, and the response of each mode can be represented using a SDOF system.
- Some common forms of dynamic loading can be expressed as a summation of several harmonic components.

Resonance 211

This section summarises the fundamental characteristics of the response of a SDOF system to a harmonic load. Many reference books, such as [16.1, 16.2], deal with other types of dynamic load as well as with multi-degrees-of-freedom systems.

16.2.1 A SDOF system subjected to a harmonic load

16.2.1.1 Equation of motion and its solution

Consider a SDOF system comprising a mass, m, attached to a spring of stiffness k, and a viscous damper of capacity c as shown in Figure 16.1a subjected to a harmonic load:

$$P(t) = P_0 \sin \omega_p t \qquad (16.1)$$

where P_0 is the force magnitude and ω_p is the angular frequency of the load. A free-body diagram of the system is shown in Figure 16.1b, from which the equation of motion can be obtained using Newton's second law:

$$m\ddot{u} = F_0 \sin \omega_p t - c\dot{u} - ku \qquad (16.2a)$$

or

$$m\ddot{u} + c\dot{u} + ku = F_0 \sin \omega_p t \qquad (16.2b)$$

(a) A SDOF system subject to a harmonic load   (b) Free-body diagram

Figure 16.1 A SDOF system subject to a harmonic force.

The particular solution of equation 16.2b can be written in the following form:

$$u(t) = A \sin(\omega_p t - \theta) \qquad (16.3)$$

where A is a constant which is equal to the maximum amplitude of the motion that follows the force by θ, called the phase lag. Equation 16.3 indicates that the mass experiences simple harmonic motion at the load frequency ω_p.

Differentiating equation 16.3 in respect to u once and twice and substituting u, $\dot{u}$ and $\ddot{u}$ into equation 16.2b leads to:

$$mA\omega_p^2 \sin(\omega_p t - \theta + \pi) + cA\omega_p \sin(\omega_p t - \theta + \pi/2) + kA\sin(\omega_p t - \theta) = P_0 \sin \omega_p t \qquad (16.4)$$

      Inertial force   Damping force   Spring force   Applied force

212 Dynamics

Equation 16.4 shows that:

- The spring or elastic force lags the applied harmonic force by an angle of θ.
- The damping force precedes the spring force by $\pi/2$.
- The inertial force precedes the damping force by $\pi/2$, but has an opposite direction to the spring force.

A vector diagram of all the forces acting on the body is shown in Figure 16.2 from which it can be shown that:

$$P_0^2 = (kA - m\omega_p^2 A)^2 + (c\omega_p A)^2 \qquad (16.5)$$

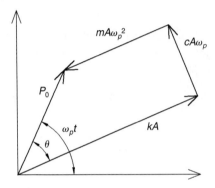

Figure 16.2 A diagram of force vectors.

The magnitude of the motion A can be obtained from equation 16.5 as:

$$A = \frac{P_0}{\sqrt{(k - m\omega_p^2)^2 + (c\omega_p)^2}}$$

$$= \frac{P_0}{k} \frac{1}{\sqrt{(1 - \beta^2)^2 + (2\xi\beta)^2}} \qquad (16.6)$$

where

$$\beta = \frac{\omega_p}{\omega} \qquad (16.7)$$

$$c = 2\xi m\omega \qquad (16.8)$$

β, called the frequency ratio, is the ratio of the frequency of the load, ω_p, to the natural frequency of the system, ω. The phase lag between the applied force and the displacement can also be found from Figure 16.2 as:

$$\tan\theta = \frac{cA\omega_p}{kA - mA\omega_p^2} = \frac{2\xi\omega_p/\omega}{1 - (\omega_p/\omega)^2} = \frac{2\xi\beta}{1 - \beta^2} \qquad (16.9)$$

Equations 16.3, 16.6 and 16.9 are the solutions of a SDOF system subject to a harmonic force in steady state vibration. They are the same as those derived from the solution of the differential equation (equation 16.2) [16.1, 16.2]. The derivation and solution indicate that:

- The displacement of the mass lags the force by θ.
- The velocity of the mass is ahead of its displacement by $\pi/2$.
- The acceleration of the mass leads its velocity by $\pi/2$ and the displacement by π.
- When $1-\beta^2=0$ or $\theta=90°$ in equation 16.9, the force diagram in Figure 16.2 becomes a rectangle. The inertial force and the spring force balance each other and the external load is countered by the damping force. As the damping force is generally much less than the inertia and spring forces, the system response will be much bigger than when the other system forces dominate.

16.2.1.2 Dynamic magnification factor

Considering the static displacement of the mass as:

$$\Delta = \frac{P_0}{k} \tag{16.10}$$

Equation 16.6 can be written in the following form:

$$A = \Delta \times M_D \tag{16.11}$$

where

$$M_D = \frac{1}{\sqrt{(1-\beta^2)^2 + (2\xi\beta)^2}} \tag{16.12}$$

Equation 16.11 indicates that *the maximum dynamic displacement (A) can be expressed as a product of the corresponding static displacement (Δ) and a dynamic magnification factor (M_D), which is thus a function of the frequency ratio and the damping ratio*. M_D therefore refers the dynamic characteristics of the system subject to a harmonic force.

For several damping ratios ($\xi=0.01$, 0.1, 0.2 and 0.5), M_D is shown in Figure 16.3 as a function of the frequency ratio (β). These curves can be interpreted as follows:

- when the load frequency is less than one-quarter the natural frequency of the system, the maximum dynamic displacement is close to the static displacement of the system;
- when the load frequency is more than twice the natural frequency, the maximum dynamic displacement is much lower than the static displacement of the system;
- when the load frequency is equal or close to the natural frequency, resonance will occur and the maximum dynamic displacement can be several times the static displacement of the system.

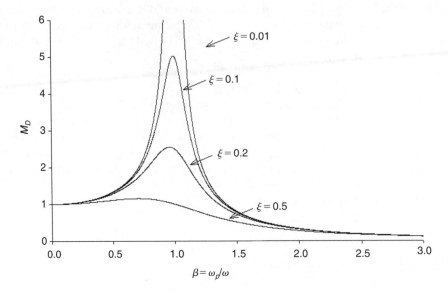

Figure 16.3 Variation of dynamic magnification factor with frequency ratio.

The significance of these observations is:

- if the load moves slowly, or the frequency of the load is much lower than the natural frequency of a structure, it can be treated as a static problem;
- if the load frequency is significantly larger than the natural frequency of a structure, dynamic analysis may not be required;
- the situation when the load frequency is close to the natural frequency of the structure will generate a resonant response and this is a situation to avoid. Increasing system damping will, however, effectively reduce the response at resonance.

16.2.1.3 The phase lag

The particular solution of equation 16.2b can be obtained by solving the differential equation and is in the following form:

$$u(t) = \frac{P_0}{k}\left[\frac{1}{(1-\beta^2)^2+(2\xi\beta)^2}\right][(1-\beta^2)\sin\omega_p t - 2\xi\beta\cos\omega_p t] \tag{16.13}$$

Let:

$$\sin\theta = \frac{2\xi\beta}{\sqrt{(1-\beta^2)^2+(2\xi\beta)^2}} \tag{16.14a}$$

$$\cos\theta = \frac{1-\beta^2}{\sqrt{(1-\beta^2)^2+(2\xi\beta)^2}} \tag{16.14b}$$

Using the relationship:

$$\sin\omega_p t\cos\theta - \cos\omega_p t\sin\theta = \sin(\omega_p t - \theta) \qquad (16.15)$$

and substituting equations 16.10, 16.12 and 16.14 into equation 16.13 leads to a concise expression:

$$u(t) = \Delta \cdot M_D \sin(\omega_p t - \theta) \qquad (16.16)$$

where θ and M_D have been given in equations 16.9 and 16.12 respectively. Equation 16.16 is normally used instead of equation 16.13 in calculations because it has a simpler form.

It should be noted that equation 16.13 and equation 16.16 do not always give identical results. The difference between equation 16.13 and equation 16.16 is due to the definitions of the angle θ in equation 16.9 and equation 16.14. Since $2\xi\beta$ will always be positive, the variation of the phase lag should be between 0 and π in equation 16.14 as shown in Figure 16.4a. However, the angle θ in equation 16.16 defined by equation 16.9 varies between $-\pi/2$ and $\pi/2$ shown in Figure 16.4b, which does not match that in the original equation (equation 16.13) as shown in Figure 16.4a. This can be examined in detail as shown in Table 16.1.

This difference is important in certain situations.

Table 16.1 Comparison of the phase lags defined in Figure 16.4a (equation 16.14) and Figure 16.4b (equation 16.9)

Situation	θ in Figure 16.4a	θ in Figure 16.4b	Conclusion
when $1 - \beta^2 > 0$	$0 < \theta < \pi/2$	$0 < \theta < \pi/2$	Equation 16.14 = Equation 16.9
when $1 - \beta^2 < 0$	$\pi/2 < \theta < \pi$	$-\pi/2 < \theta < 0$	Equation 16.14 ≠ Equation 16.9

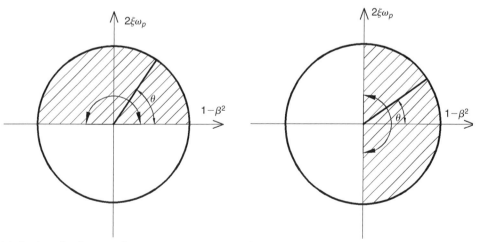

(a) $0 < \theta < \pi$ in the actual situation (Equation 16.14)

(b) $-\pi/2 < \theta < \pi/2$ in Equation 16.9

Figure 16.4 Definition of the range of the phase lag, θ.

Example 16.1

A SDOF system subjected to a harmonic load has the following parameters:

$$F_0 = 1\,\text{N};\ k = 1\,\text{kN/m};\ \beta = 1.1;\ \xi = 0.02;\ \omega_p = 2\pi\,\text{rad/s}$$

Calculate the response of the system when the phase lags are calculated using equation 16.9 and equation 16.14 respectively.

Solution

Substituting the above data into equation 16.14 gives:

$$2\xi\beta = 0.044 \quad 1-\beta^2 = -0.21 \quad \sqrt{(1-\beta^2)^2 + (2\xi\beta)^2} = 0.2146$$

$$\sin\theta = 0.2050 \quad \theta \text{ can be } 11.83° \text{ or } 168.17°$$

$$\cos\theta = -0.9787 \quad \theta \text{ can be } 168.17° \text{ or } -168.17°$$

The actual phase lag should satisfy both equations 16.14a and 16.14b. Thus it must be 168.17°.

When equation 16.9 is used, it yields

$$\tan\theta = -0.2095 \quad \text{and} \quad \theta = -11.83°$$

The phase difference between −11.83° an 168.17° is exactly 180°. The response curves generated using equation 16.13 and equation 16.16 are given in Figure 16.5 showing the 180° phase difference.

To this incompatibility, a supplementary condition must be introduced and equation 16.9 should be written in the following form:

$$\theta = \begin{cases} \tan^{-1}\left[\dfrac{2\xi\beta}{1-\beta^2}\right] + \pi & \text{if } 1-\beta^2 < 0 \\[2ex] \tan^{-1}\left[\dfrac{2\xi\beta}{1-\beta^2}\right] & \text{if } 1-\beta^2 > 0 \end{cases} \tag{16.17}$$

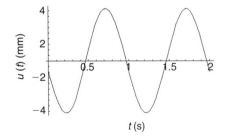

(a) Displacements calculated using Equation 16.13

(b) Displacements calculated using Equation 16.16

Figure 16.5 Harmonic vibration of a damped SDOF system.

Then the response calculated using equations 16.16 and 16.17 is the same as that using equations 16.13 and 16.14.

The supplementary condition (equation 16.17) is necessary and straightforward when equation 16.16 is used, but this has not been emphasised elsewhere. The possible error in calculating phase lag does not affect the magnitude of the response when a SDOF system is subjected to a single harmonic load; however, when several harmonics, such as those for human jumping loads, need to be considered, the errors in the calculation of phase lags will affect the magnitudes and pattern of the loads and consequently the response of the structure subjected to these loads [16.3].

16.2.2 A SDOF system subject to a harmonic support movement

Consider a SDOF system subjected to vertical harmonic ground or support movements. The ground motion is defined as follows:

$$v_g = B\sin\omega_p t \tag{16.18}$$

where B is the magnitude of the ground movement and ω_p is the frequency of the ground movement.

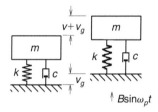

Figure 16.6 A SDOF system subject to vertical ground movements.

The equation of motion of the SDOF system is:

$$m(\ddot{v}_g + \ddot{v}) + c\dot{v} + kv = 0 \tag{16.19}$$

where v is the movement of the mass relative to the ground. Substituting equation 16.18 into equation 16.19 gives:

$$m\ddot{v} + c\dot{v} + kv = m\omega_p^2 B\sin\omega_p t \tag{16.20}$$

Comparing equation 16.20 and equation 16.2b, it can be seen that they are identical when $P_0 = Bm\omega_p^2$. Thus the solution of equation 16.20 has the same form as equation 16.3 and the magnitude of the vibration using equation 16.6 is:

$$A = \frac{Bm\omega_p^2}{\sqrt{(k - m\omega_p^2)^2 + (c\omega_p)^2}}$$

$$= \frac{Bm\omega_p^2}{m\omega^2} \frac{1}{\sqrt{(1 - \omega_p^2/\omega^2)^2 + (2\xi\omega_p/\omega)^2}} = \frac{B\beta^2}{\sqrt{(1 - \beta^2)^2 + (2\xi\beta)^2}} \tag{16.21}$$

Thus the solution of equation 16.20 is:

$$v(t) = BM_R \sin(\omega_p t - \theta) \tag{16.22}$$

where the phase lag θ has been given in equation 16.17 and M_R is the dynamic magnification factor for the relative movement of the system to ground and is defined as:

$$M_R = \frac{\beta^2}{\sqrt{(1-\beta^2)^2 + (2\xi\beta)^2}} \tag{16.23}$$

If the absolute motion, $v_a = v_g + v$, of the mass is considered, equation 16.19 can be rewritten as:

$$m\ddot{v}_a + c\dot{v}_a + kv_a = c\dot{v}_g + kv_g$$

$$= cB\omega_p \cos\omega_p t + kB\sin\omega_p t$$

$$= B\sqrt{k^2 + (c\omega_p)^2}\,\sin(\omega_p t + \phi) \tag{16.24}$$

where

$$\phi = \tan^{-1}\left[\frac{c\omega_p}{k}\right] = \tan^{-1}[2\xi\omega_p/\omega] \tag{16.25}$$

As both $2\xi\omega_p$ and ω are positive, ϕ varies between 0 and $\pi/2$. Thus, similar to equation 16.22, the solution of equation 16.24 can be written as:

$$v_a(t) = BM_A \sin(\omega_p t + \phi - \theta) \tag{16.26}$$

where M_A is the dynamic magnification factor for the absolute movement of the system and is expressed as:

$$M_A = \frac{\sqrt{1 + (2\xi\beta)^2}}{\sqrt{(1-\beta^2)^2 + (2\xi\beta)^2}} \tag{16.27}$$

The phase lag in equation 16.26 has been defined in equation 16.17.

Comparing the ground motion, equation 16.18, and the absolute movement of the system, equation 16.26, it can be noted that the magnification factor M_A in equation 16.27 is also called the motion transmissibility that is the ratio of the amplitude of the absolute vibration of the system to the amplitude of the ground motion. Similar to Figure 16.3 for the magnification factor when the system is subjected to a harmonic load, Figures 16.7 and 16.8 show the dynamic magnification factors for the relative and absolute displacement respectively for four damping values, 0.01, 0.1, 0.2 and 0.5 when the system is subjected to harmonic ground motion.

The same qualitative observations as those from Figure 16.3 can be obtained from Figure 16.8.

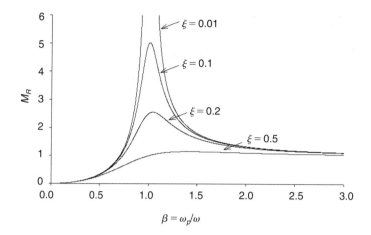

Figure 16.7 The magnification factor for the relative movement of the system.

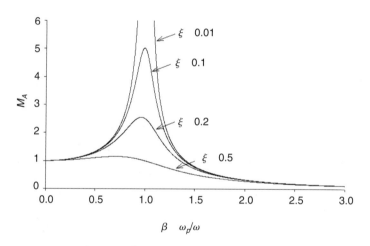

Figure 16.8 The magnification factor for the absolute movement (motion transmissibility) of the system.

16.2.3 Resonance frequency

It can be observed from Figures 16.3, 16.7 and 16.8 that the resonance frequency is different to and maybe larger or smaller than the natural frequency of the system. This difference is more pronounced for higher damping ratios. The resonance frequency can be obtained by differentiating the magnification factors with respect to the frequency ratio and letting the functions be zero. For the magnification factors given in equations 16.12, 16.23 and 16.27:

$$\frac{\partial M_D}{\partial \beta} = \frac{\partial}{\partial \beta}\left(\frac{1}{\sqrt{(1-\beta^2)^2+(2\xi\beta)^2}}\right)=0 \qquad (16.28a)$$

$$\frac{\partial M_R}{\partial \beta} = \frac{\partial}{\partial \beta}\left(\frac{\beta^2}{\sqrt{(1-\beta^2)^2+(2\xi\beta)^2}}\right)=0 \qquad (16.28b)$$

$$\frac{\partial M_A}{\partial \beta} = \frac{\partial}{\partial \beta}\left(\frac{\sqrt{1+(2\xi\beta)^2}}{\sqrt{(1-\beta^2)^2+(2\xi\beta)^2}}\right)=0 \qquad (16.28c)$$

The solution of equation 16.28 gives the relationship between the resonance and natural frequencies of a SDOF system as follows:
When subjected to a harmonic load:

$$f_R = f\sqrt{1-2\xi^2} \qquad (16.29a)$$

When subjected to a harmonic ground motion and examining the relative movement:

$$f_{R,R} = \frac{f}{\sqrt{1-2\xi^2}} \qquad (16.29b)$$

When subjected to a harmonic ground motion and considering the absolute movement:

$$f_{R,A} = \sqrt{\frac{\sqrt{1+8\xi^2}-1}{4\xi^2}} \qquad (16.29c)$$

Thus the differences between the natural frequency and the three resonance frequencies can be quantified using equation 16.29. Table 16.2 lists the ratios of the resonance frequency to the natural frequency for different damping ratios and for the three expressions.

Table 16.2 Effect of damping ratio on the resonance frequency

	ξ	0.01	0.05	0.1	0.2	0.3	0.4	0.5
Equation 16.29a	f_R/f	0.9999	0.9975	0.9899	0.9592	0.9055	0.8246	0.7071
Equation 16.29b	$f_{R,R}/f$	1.0001	1.0025	1.0102	1.0426	1.1043	1.2127	1.4142
Equation 16.29c	$f_{R,A}/f$	0.9999	0.9975	0.9903	0.9647	0.9301	0.8927	0.8556

It can be observed from Table 16.2 that:

- The resonance frequency is related to the damping ratio. When the damping ratio is less than 0.1, there are negligible differences between the natural frequency and resonance frequency.
- The resonance frequency is also related to the input loading or ground motion and to the type of displacement (relative or absolute movement).

Structures in civil engineering normally have damping ratios smaller than 10 per cent. Therefore there is usually no need to distinguish between the resonance frequency and natural frequency in the vibration of structures.

16.3 Model demonstrations

To see a resonant response, it requires an input device that can generate either harmonic base motion, such as a shaking table, or harmonic load, such as using a vibrator, together with a simple structure that has a natural frequency within the frequency range of the input.

16.3.1 Dynamic response of a SDOF system subject to harmonic support movements

This demonstration shows *the observations obtained from Figures 16.3 and 16.8, i.e. the relationship between the response of a SDOF system and the ratio of the frequency of input to the natural frequency of the system.*

Figure 16.9 The model of a SDOF system.

An elastic string and a mass can form a simple SDOF system as shown in Figure 16.9. Hold the end of the string and move it up and down harmonically to simulate the harmonic support movement of the SDOF system. This corresponds to the model shown in Figure 16.6. The movement of the mass can be described using equation 16.24, and the ratio of the maximum movement of the mass to that of the hand is shown in Figure 16.8.

The dynamic phenomena shown in Figure 16.8 can be demonstrated qualitatively as follows:

- The person holding the string first moves his/her hand slowly up and down, creating a situation where the frequency of the support (hand) movement is much smaller than the natural frequency of the SDOF system. It will be observed that the amplitude of movement of the mass is almost the same as that of the support (hand), as indicated by Figure 16.8 when the frequency ratio is less than 0.25.
- When the hand moves up and down quickly, it creates a situation that the frequency of the hand movement is larger than the natural frequency of the system.

It can be seen that the hand movements are much larger than the movements of the mass, as indicated in Figure 16.8 when the frequency ratio is larger than 2.0.
- Finally, when the hand moves up and down at a frequency close to the natural frequency of the system, a situation is created in which resonance develops. It is observed that the movements of the mass are much larger than the hand movements, as indicated in Figure 16.8 when the frequency ratio is near to unity.

16.3.2 Effect of resonance

This demonstration shows that *a cantilever experiences significant vibration at resonance*.

Figure 16.10a shows a medical shaker that can generate vibrations at a frequency of 5 Hz or 10 Hz in three perpendicular directions. A steel ruler and a longer and thicker steel ruler are mounted on a wooden plate that can be firmly placed on the shaker as shown in Figure 16.10b. Both rulers are selected to have their fundamental natural frequencies slightly less than 5 Hz.

Switch on the shaker at the frequency of 10 Hz and it will be seen that the two rulers vibrate with very small amplitudes in comparison with the movement of the shaker.

Change the vibration frequency from 10 Hz to 5 Hz and it will then be observed that both rulers vibrate with significant amplitudes as the fundamental frequencies of the rulers are close to the shaking frequency.

(a) A medical shaker and two cantilevers (b) Simulated shaking table test

Figure 16.10 Demonstration of resonance.

16.4 Practical examples

In certain situations resonance should be avoided in civil engineering structures. This requires the knowledge of the natural frequencies of the structure and the frequencies of the dynamic loading applied to it.

16.4.1 The London Millennium Footbridge

The London Millennium Footbridge is the first new pedestrian bridge crossing over the Thames in Central London for more than a century. The bridge is located between Peter's Hall on the north bank leading to Saint Paul's Cathedral and the new Tate Modern Art Gallery on the south bank. The bridge has three spans over a total length of 325 metres. The lengths of the three spans are 81 metres for the north span, 144 metres for the main span and 100 metres for the south span.

Figure 16.11 The London Millennium Footbridge.

The London Millennium Footbridge, Figure 16.11, opened on 10 June 2000. It was estimated that between 80000 and 100000 people crossed the bridge during the opening day with a maximum of 2000 people on the bridge at any one time. During the opening day unexpected lateral movements, or 'wobbling', occurred when people walked across the bridge. Two days later the bridge was officially closed in order to investigate the causes of the vibration. After investigation and modification, the bridge reopened in December 2001 [16.4]. Since then millions of people have visited the bridge and there has been no reoccurrence of 'wobbling' problems.

As constructed, the bridge had lateral natural frequencies as follows:

- south span: first lateral natural frequency was about 0.8 Hz;
- central span: first two lateral natural frequencies were about 0.5 and 0.95 Hz;
- north span: first lateral natural frequency was about 1 Hz.

It was reported that excessive vibrations of the bridge in the lateral direction did not occur continuously but built up when a large number of pedestrians were on the south and central spans of the bridge and then died down if the number of people on the bridge reduced or if the people stopped walking [16.4].

Resonance is related to forcing frequencies. Walking is a periodic movement on a flat surface in which two feet move alternately from one position to another and do

224 Dynamics

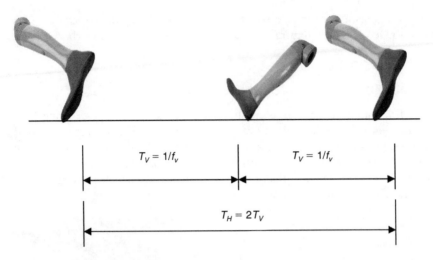

Figure 16.12 Periods of walking forces.

not leave the surface simultaneously. When people walk in a normal way, walking has its own frequency. This is illustrated in Figure 16.12.

When people walk, they produce vertical loading on the walking surface, such as floors and footbridges. The time from one vertical load generated by one foot to the next vertical load induced by the other foot is the period of the walking load in the vertical direction, denoted as T_V. People walking generate not only vertical forces, but also lateral forces. The force generated laterally by the right foot is normally in the opposite direction to the force induced by the left foot. Therefore the time required for reproducing the same pattern of force can be counted from the left (or right) foot to the next step of the left (or right) foot. In other words, the period of lateral forces (T_H) is just twice the period of the vertical force induced by people walking, or the frequency of the lateral walking loads is a half of that of the vertical walking loads [16.5].

Figure 16.13 shows the distribution of walking frequencies obtained from 400

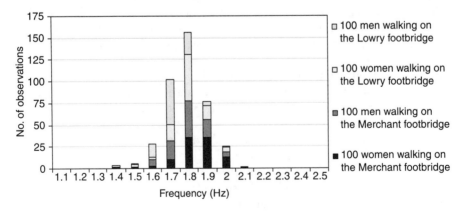

Figure 16.13 Distribution of frequency of walking loads in the vertical direction [16.6].

people walking on two footbridges in Manchester [16.6]. The horizontal axis indicates the walking frequency in the vertical direction while the vertical axis shows the number of observations of particular frequencies. It can be seen from Figure 16.13 that most of the 400 frequencies of people walking are between 1.6 Hz and 2 Hz in the vertical direction. As the walking frequency in the lateral direction is just half that in the vertical direction, the corresponding frequencies of people walking in the lateral direction are between 0.8 Hz and 1 Hz.

It can be noted that the frequencies of walking loads in the lateral directions were close to the lateral natural frequencies of the south and central spans of the London Millennium Footbridge where excessive vibrations occurred. It was also observed that people walking in large groups tended to synchronise their walking paces. When the footbridge started wobbling, more people would walk at the frequency of the wobbling, which enhanced the synchronisation. This synchronisation magnified the effect of the lateral footfall forces on the footbridge. The wobbling of the footbridge was caused by resonance that synchronisated the walking loads.

16.4.2 Avoidance of resonance: design of structures used for pop concerts

British Standard BS 6399: Part 1, Loading for Buildings [16.6], introduced in September 1996, included a new section on synchronised dance loading. It stated that any structure that might be subjected to this form of loading should be designed in one of two ways: to withstand the anticipated dynamic loads or to avoid significant resonance effects. The first method requires dynamic analysis to assess the structural response to the loading in order to calculate a safety margin. The second method requires that the structure should be designed to be sufficiently stiff so that the lowest relevant natural frequency of the structure is above the range of load frequencies considered. The second approach is simpler and only requires the calculation of natural frequencies. However, both approaches need knowledge of the load frequencies.

Dance-type activities, such as keep-fit exercises, aerobics and audience movements at pop concerts, are more common now than ever before. These activities are likely to be held on grandstands, dance floors and in sports centres. Therefore, the use and/or design of these structures should consider the effect of the human-induced dance-type loads. Figure 16.14 shows a pop concert where people moved, jumped, bobbed and swayed, in time to the music.

Dance is movement with rhythmic steps and actions, usually to accompanying music. Efforts have been made to study dance-type loads since prediction of the response of a structure subject to this loading requires an understanding of the loading. Dance-type loads are functions of the type of dance activity, the density and distribution of the dancers, the frequency of the music, load factors and the dynamic crowd effect.

There are many different types of dancing and a wide range of beat frequencies for dance music, however, dance frequencies tend to be in the range of 1.5–3.5 Hz for individuals and 1.5–2.8 Hz for groups of people.

Figure 16.15 shows an autospectrum obtained from accelerometers monitoring vertical motions of a grandstand during a pop concert [16.8]. The vertical axis indicates the normalised acceleration squared per Hz, while the horizontal axis shows

Figure 16.14 At a pop concert.

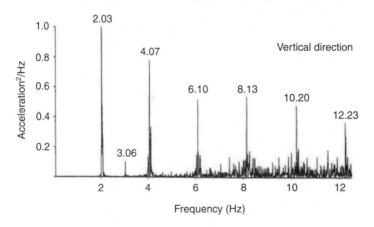

Figure 16.15 Structural response to the music beat frequency at a pop concert [16.8].

frequency. It can be seen that the frequencies corresponding to the peaks of responses are the beat frequency of the music and integer multiples of the beat frequency. This phenomenon has been observed in a number of measurements taken during pop concerts and can be explained theoretically. As spectators at pop concerts move with the music beat, the frequency of dance-type loads can be determined from the beat frequency of music played on these occasions. The beat frequencies of 210 modern songs have been determined [16.9].

The 210 songs consisted of 30 songs from each of the 1960s, the 1970s and the 1980s, and 120 songs from the 1990s. The 1990s music was further classified into four main types, dance, indie, pop and rock, with 30 songs for each group. All the songs selected had been popular in their time and thus provided a good sample of popular music. The frequency distribution of the 210 songs surveyed is given in Figure 16.16. It can be seen that the majority of songs (202 out of 210) are in the frequency range of 1–2.8 Hz.

Human loading, such as jumping, bobbing and walking, contains several harmonic components with the load frequency, two times the frequency, three times the frequency and so on. However, only the first two or three components are significant and need to be considered in design. For example, if the load or music frequency is not larger than 2.8 Hz and the first three load components are considered for a floor subject to jumping type loading, the highest frequency in the load to be considered will not be larger than $2.8 \times 3 = 8.4$ Hz. This corresponds to the threshold value of natural frequency in BS6399: Part 1 of 8.4 Hz in the vertical direction, for which no dynamic analysis of the structure is required [16.7].

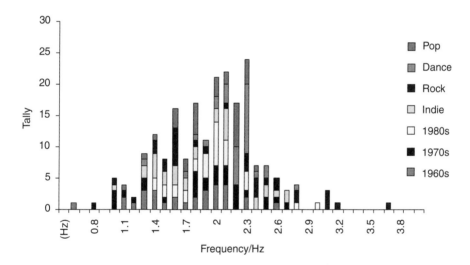

Figure 16.16 Frequency distribution of the 210 modern songs [16.9].

16.4.3 Measurement of the resonance frequency of a building

The concept of resonance can be used to identify the natural frequencies of a structure. A harmonic force with appropriate amplitude is applied to the structure and the corresponding maximum structural response is then recorded. The procedure is repeated a number of times with different forcing frequencies. The frequency corresponding to the largest value of these maximum responses is the resonant frequency of the structure. As civil engineering structures have damping ratios far less than 10 per cent, the resonance frequency measured from the forced vibration is actually the natural frequency of the structure, as shown in Table 16.2.

228 *Dynamics*

Among a number of experiments carried out on the test building shown in Figure 15.17 were forced vibration tests with vibration generators used to shake the structure in a controlled manner at frequencies within the range 0.3 Hz to 20 Hz [16.10]. Four vibration generators were placed at the four corners of the roof of the building and the building response was monitored using accelerometers aligned to the measure motions in appropriate directions. The response of the building was sampled using optimised filtering, amplification and curve fitting and then normalised by converting the measured accelerations to equivalent displacements which were then divided by the applied forces. The maximum normalised displacement corresponding to a particular load frequency was plotted as a cross in Figure 16.17. This process was repeated for a number of load frequencies to produce many crosses in Figure 16.17, which are linked by a best-fit one-degree-of-freedom curve. The frequency corresponding to the largest response is a resonance frequency of the structure. For this building the resonant frequency was 0.617 Hz in one main direction. This technique has been widely used in structural engineering, mechanical engineering and other areas.

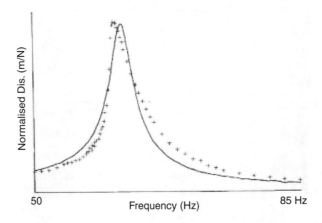

Figure 16.17 Frequency spectrum in one main direction of the building shown in Figure 15.17 [16.10].

16.4.4 An entertaining resonance phenomenon

Figure 16.18a shows a replica bronze washbowl which was used in ancient China. Four fish are cast into the bottom of the washbowl in an anticlockwise direction around the centre of the base as shown in Figure 16.18b. The washbowl is symmetric about the plane that passes through the centre of the bowl and is parallel to the two handles. Now, such bowls are used for entertainment, as the water in the washbowl can be made to spurt up to 300 mm into the air around the locations of the fishes' mouths when two hands rub the handles on the washbowl. This effect can be produced as follows:

1 The washbowl is filled half full with water and placed on a wet kitchen towel set in the shape of a ring of radius of about 80 mm.
2 Clean two hands using soap and water and the palms are used to rub the

handles of the washbowl as shown in Figure 16.18c. The two palms move forward and backward alternatively and periodically, creating a pair of antisymmetric periodical forces applied to the symmetrically located handles of the washbowl.
3   Sound will generate from between the palms and the handles and ripples will form on the water surface (Figure 16.18c).
4   Continuing the hand movements will cause drops of water to jump out almost vertically from locations around the mouths of the fishes. Figure 16.18d shows the situation when the resonance occurs.

Only small forces are applied to the two handles of the washbowl in the vertical and forward–backward directions. As the hands move periodically, resonance occurs when the frequency of hand movement matches one of the natural frequencies of the washbowl–water system. The evidence of the resonance is sound and 'jumping water'.

(a) Replica bronze washbowl   (b) Four fishes cast into the bottom of the bowl

(c) Ripple on water surface   (d) Water jumping out vertically

Figure 16.18 An entertaining resonance phenomenon.

References

16.1 Beards, C. F. (1996) *Structural Vibration: Analysis and Damping*, London: Arnold.

16.2 Clough, R. W. and Penzien, J. (1993) *Dynamics of Structures*, New York: McGraw-Hill.

16.3 Ji, T. and Wang, D. (2001) 'A supplementary condition for calculating periodical vibration', *Journal of Sound and Vibration*, Vol. 241, No. 5, pp. 920–924.

16.4 Dallard, P., Fitzpatrick, A. J., Flint, A., Le Bourva, S., Low, A., Ridsdill-Smith, R. M. and Willford, M. (2001) 'The London Millennium Footbridge', *The Structural Engineer*, Vol. 79, pp. 17–33.

16.5 Ellis, B. R. (2001) 'Serviceability evaluation of floor vibration induced by walking loads', *The Structural Engineer*, Vol. 79, No. 21, pp. 30–36.

16.6 Pachi, A. and Ji, T. (2005) 'Frequency and velocity of walking people', *The Structural Engineer*, Vol. 83, No. 3, pp. 36–40.

16.7 BSI (1996) BS 6399: Part 1: Loading for Buildings, London.

16.8 Littler, J. D. (1998) 'Full-scale testing of large cantilever grandstands to determine their dynamic response', *Proceedings of the First International Conference on Stadia 2000*, pp. 123–134.

16.9 Ginty, D., Derwent, J. M. and Ji, T. (2001) 'The frequency ranges of dance-type loads', *Journal of Structural Engineer*, Vol. 79, No. 6, pp. 27–31.

16.10 Ellis, B. R. and Ji, T. (1996) 'Dynamic testing and numerical modelling of the Cardington steel framed building from construction to completion', *The Structural Engineer*, Vol. 74, No. 11, pp. 186–192.

17 Damping in structures

17.1 Concepts

- The larger the damping ratio, ξ, the larger the ratio of successive peak displacements in free vibration (u_n/u_{n+1}), and the quicker the decay of oscillations.
- The damping ratio can be determined from measurements using the vibration theory of a single-degree-of-freedom system.
- The damping ratio is a measure of the amount of damping in a structure which can effectively reduce structural vibration at resonance.
- The higher the amplitude of free vibration of a structure, the larger will be the critical damping ratio and the smaller will be the natural frequency.

17.2 Theoretical background

The process by which vibrations decrease in amplitude with time is called damping. Damping dissipates the energy of a vibrating system and can do this in a number of ways. There are several different types of damping: friction, hysteretic and viscous. For structural systems the damping encountered is best described using a viscous model in which the damping force is proportional to its velocity. The damping ratio ξ is expressed as a fraction of critical damping and is often written as a percentage (i.e. 1 per cent critical).

The value of the damping ratio is required for predicting structural responses induced by different forms of dynamic loading. When structural vibrations are too large, artificial damping devices can be used to increase the damping and therefore reduce the levels of the vibrations. For example, dampers were installed in the London Millennium Footbridge to reduce the large lateral vibration of the bridge that occurred as a result of people walking across the bridge.

17.2.1 Evaluation of viscous-damping ratio from free vibration tests

Consider the equation of motion of a single-degree-of-freedom system with viscous damping:

$$m\ddot{u} + c\dot{u} + ku = 0 \tag{17.1}$$

The solution of equation 17.1 has been given in equation 15.15 as follows:

$$u(t) = \left[\frac{\dot{u}(0) + u(0)\xi\omega}{\omega_D}\sin\omega_D t + u(0)\cos\omega_D t\right]e^{-\xi\omega t} \tag{15.15}$$

232 Dynamics

where ω and ω_D are the angular frequencies of the undamped and damped systems respectively.

Example 17.1

Consider free vibrations of an undercritically damped system that has the following properties: $f = \omega/2\pi = 1.0\,\text{Hz}$, $\xi = 0.05$, $u(0) = 10\,\text{mm}$ and $\dot{u}(0) = 0$

Solution

Substituting the above data into equation 15.15, the displacements of the system can be represented in graphical form as shown in Figure 17.1.

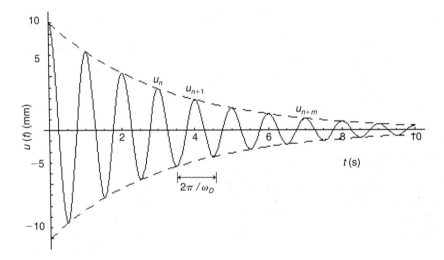

Figure 17.1 Damped free vibration.

It can be seen that:

- the damped system oscillates about its neutral position with a constant angular frequency ω_D;
- the oscillation decays exponentially due to the damping.

Consider any two successive positive peaks such as u_n and u_{n+1} which occur at time $n(2\pi/\omega_D)$ and $(n+1)(2\pi/\omega_D)$ respectively. Equation 15.15 can be used to obtain the ratio of the two successive peak values as:

$$\frac{u_n}{u_{n+1}} = \exp(2\pi\xi\omega/\omega_D) \tag{17.2}$$

Taking the natural logarithm of both sides of the equation and substituting $\omega_D = \omega\sqrt{1-\xi^2}$ into equation 17.2, the logarithmic decrement of damping, δ, is obtained since ξ is normally small for structural systems:

$$\delta = \ln \frac{u_n}{u_{n+1}} = \frac{2\pi\xi}{\sqrt{1-\xi^2}} = 2\pi\xi \quad \text{or} \quad \xi = \frac{\delta}{2\pi} = \frac{1}{2\pi}\ln\frac{u_n}{u_{n+1}} \tag{17.3}$$

Equation 17.3 indicates that:
The larger the damping ratio ξ, the larger the ratio of u_n/u_{n+1} and thus the quicker the decay of the oscillation.

Considering two positive peaks from Figure 17.1 which are not adjacent, say u_n and u_{n+m}, equation 17.3 becomes:

$$\delta = \ln \frac{u_n}{u_{n+m}} = \frac{2\pi m\xi}{\sqrt{1-\xi^2}} = 2\pi m\xi \quad \text{or} \quad \xi = \frac{\delta}{2m\pi} \tag{17.4}$$

Thus the damping ratio can be evaluated from equation 17.3 or equation 17.4 by measuring any two positive peak displacements and the number of cycles between them.

If the damping is perfectly viscous, equation 17.3 and equation 17.4 will give the same result. However, this may not be true in all practical situations as will be shown in section 17.4.

The advantages of using free vibration test responses to obtain values of damping ratio are:

- the requirements for equipment and instrumentation are minimal in comparison with forced vibration test methods;
- only relative displacement amplitudes need to be measured;
- the initial vibration can be generated by any convenient method, such as an initial displacement, an impulse or a sudden change of motion.

17.2.2 Evaluation of viscous-damping ratio from forced vibration tests

The dynamic magnification factor of a single-degree-of-freedom system subject to a harmonic load is shown in section 16.2, i.e.:

$$M_D = \frac{A}{\Delta_{st}} = \frac{1}{\sqrt{(1-f_p^2/f^2)^2 + (2\xi f_p/f)^2}} \tag{16.12}$$

where f_p is the frequency of the load and f is the natural frequency of the system. At resonance of a structure with a low damping ratio, the frequency of oscillation is approximately equal to the natural frequency of the structure, i.e. $f_p = f$ and equation 16.12 becomes:

$$M_D = \frac{A}{\Delta_{st}} = \frac{1}{2\xi} \tag{17.5}$$

Thus if both the static displacement and the maximum dynamic displacement of a single-degree-of-freedom system (or a particular mode of a structure) at resonance can be determined experimentally, the critical damping ratio can be calculated from equation 17.5. More accurate methods for determining the damping ratio from forced vibration tests can be found in references [17.1, 17.2 and 17.3].

234 *Dynamics*

17.3 Model demonstrations

17.3.1 Observing the effect of damping in free vibrations

This set of models demonstrates *the effect of damping provided by oil in free vibration, which can be observed by eyes.*

Take a steel ruler and form a cantilever as shown in Figure 15.9a. Give the tip of the ruler an initial displacement and release it. The steel ruler will perform many cycles of oscillation before it becomes stationary, indicating that the damping ratio associated with steel alone is low.

Figure 17.2a shows two identical steel strips acting as cantilevers. The only difference between the two cantilevers is that connected to the free end of the cantilever on the left is a vertical metal bar which in turn is attached to a disk immersed in oil. Figure 17.2b shows the disk by lifting up the free end of the strip. Thus the effect of the oil and the device can be examined.

Press the free ends of the two cantilevers down by the same amount and then release them suddenly. It will be observed that the cantilever on the left only vibrates for a small number of cycles before stopping while the cantilever on the right oscillates through many cycles demonstrating the effect of viscous damping provided by the oil.

17.3.2 Hearing the effect of damping in free vibrations

This set of models shows *the effect of damping provided by rubber bands in free vibration, which can be heard by ears.*

(a)

(b)

Figure 17.2 Effect of the damping provided by oil.

The sound heard from the free vibrations of a taut string, such as a violin string, links two different physical phenomena, sound transmission and the string vibrations, which both can be described using the same differential equation of motion. Thus hearing sound can be related to observing free vibrations, in this case those of a taut string.

Figure 17.3 shows two identical steel bars, one is a bare bar and the other has rubber bands wrapped around it. The effect of the damping added by the rubber can be demonstrated as follows:

- Suspend the bare bar and give it a knock at its lower end using the other metal bar as shown in Figure 17.3a. A sound will be generated from the bar for several seconds as it reverberates.
- Suspend the wrapped bar and give it a similar knock on the exposed metal part (Figure 17.3b). This time only a brief dull sound is heard as the rubber wrapping dissipates much of the energy of the vibration.

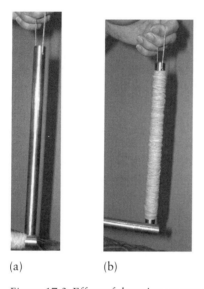

(a)   (b)

Figure 17.3 Effect of damping on sound transmission.

17.4 Practical examples

17.4.1 Damping ratio obtained from free vibration tests

The true damping characteristics of typical structural systems are normally very complex and difficult to define with few structures actually behaving as ideal single-degree-of-freedom systems. Notwithstanding this, the earlier discussion of single-degree-of-freedom systems can be useful when considering more complex practical structures.

A free vibration test was conducted on a full-sized eight-storey test building (Figure 15.19) to identify the critical damping ratio of its fundamental mode [17.4]. In order to amplify the displacements of the structure the building was shaken, by a set of vibrators mounted at the four corners on the roof of the building, at the fundamental frequency of the building. After the vibrators were suddenly stopped, the ensuing free vibrations of the roof of the building were measured. The decay of the vibrations in one of the two main directions of the building is shown in Figure 15.18. As the excitation caused vibrations effectively only in the fundamental mode, the contributions of other modes of vibration to the decaying response of the structure were negligible.

Table 17.1 Natural frequency and damping ratio determined from various sections of decay

Relative amplitude	Natural frequency (Hz)	Damping ratio (%)
1.00	0.611	2.87
0.366	0.636	1.81
0.181	0.645	1.28
0.106	0.647	1.02
0.062	0.656	0.85

The natural frequency and damping ratio of the response of the structure can be determined from the records shown in Figure 15.18. Five continuous 10s samples of vibrations were extracted from the response and a curve-fitting technique was used to produce smooth curves from which the natural frequency and damping ratio could be determined. One such smoothed curve, superimposed on the measured curve, is shown in Figure 17.4. The response frequency and damping ratio values extracted from the five samples are given in Table 17.1 and related to the amplitude of vibration at the start of each sample [17.4].

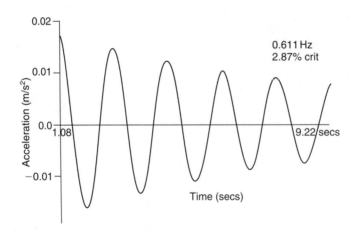

Figure 17.4 Extraction of natural frequency and damping ratio from free vibration records [17.4].

From Table 17.1 it can be observed that:

- the higher the amplitude of vibration, the smaller the natural frequency and the larger the damping ratio;

Damping in structures 237

- the natural frequency for the relative amplitude of 1.00 is 6.86 per cent lower than that for the relative amplitude of 0.062 while the damping ratio at the relative amplitude of 1.00 is 238 per cent higher than that at the relative amplitude of 0.062.

When the building vibrated with small amplitudes, the relative movements between joints and other connections in the structure were small involving frictional forces doing less work leading to lower damping ratios than were found when amplitudes were larger with associated larger relative joint movements and friction related work. These variations have been observed in many different types of structure.

17.4.2 Damping ratio obtained from forced vibration tests

The forced vibration tests of the framed steel building for obtaining its resonance frequency are described in section 16.4.3. The frequency spectrum obtained from the experiment and curve fitting are shown in Figure 16.17.

The natural frequency and damping ratio obtained from the forced vibration tests were 0.617 Hz and 2.25 per cent respectively. It can be noted that the measured values have a characteristic negative skew compared to the best-fit curve. This is typical of this type of measurement and shows one aspect of non-linear behaviour as observed in free vibration tests of buildings and other structures.

It can be observed from Table 9.1 that the measurements from the forced vibration tests agree favourably with those obtained from the free vibration tests for relative amplitudes between 0.366 and 1.000. In general, forced vibration tests normally provide larger forces than free vibration tests. Therefore forced vibration tests frequently produce smaller values for natural frequencies and larger values for damping ratios than those obtained from free vibration tests.

17.4.3 Reducing footbridge vibrations induced by walking

A total of 37 viscous dampers were installed on the London Millennium Footbridge in order to remove the lateral vibrations of the bridge which occurred when people walked across the bridge. The majority of these dampers are situated beneath the bridge deck supported by transverse members (Figure 17.5a). One end of each

(a)   (b)

Figure 17.5 Damping devices installed on the London Millennium Footbridge.

viscous damper is connected to the apex of a steel V brace, known as a chevron. The apex of the chevron is supported on roller bearings that provide vertical support but allow sliding in the other directions. The other ends of the chevron are fixed to the neighbouring transverse members [17.5].

Viscous dampers were also installed in the planes between the cables and the deck at the piers (Figure 17.5b) to provide damping of the lateral and lateral–torsional modes of vibration.

17.4.4 Reducing floor vibration induced by walking

Damping can be introduced into concrete floors by sandwiching a layer of high damping material between the structural concrete floor and a protective concrete topping. This acts in a similar manner to the demonstration model described in section 15.3.3.

During the bending of the floor induced by footfall vibrations, energy is dissipated through the shear deformation produced in the damping material by the relative deformations of the two concrete layers (Figure 17.6). The technique was originally used to damp out vibrations of aircraft fuselage panels when the resonant frequencies of the panels were over 200 Hz.

A floor panel, 6 m by 9 m, in the steel-frame test building at the BRE Cardington Laboratory was selected for testing with and without damping layers (Figure 17.7). A variety of comparative tests were conducted including heel-drop tests, forced vibration tests and walking tests at different paces.

Figure 17.8 shows the comparison of acceleration–time histories induced by the same individual walking on the floor panel without and with the damping layer. The benefit of the constrained damping layer in reducing the vibration of the floor is obvious.

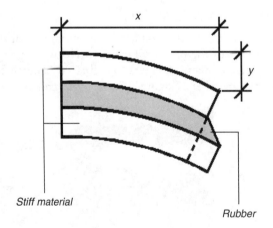

Figure 17.6 Deformation of bonded layers.

Figure 17.7 The test floor (courtesy of Dr B. Ellis).

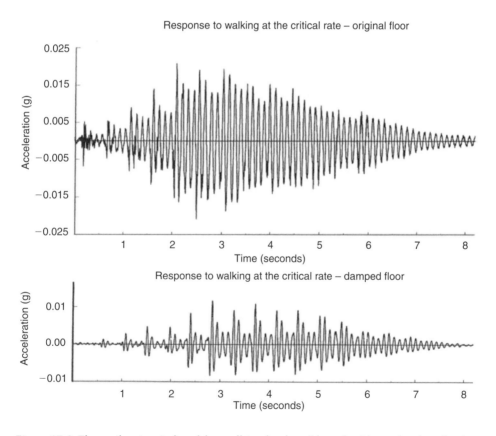

Figure 17.8 Floor vibration induced by walking loads, with and without the damping layer (courtesy of Dr B. Ellis).

References

17.1 Beards, C. F. (1996) *Structural Vibration: Analysis and Damping*, London: Arnold.
17.2 Clough, R. W. and Penzien, J. (1993) *Dynamics of Structures*, New York: McGraw-Hill.
17.3 Chopra, A. K. (1995) *Dynamics of Structures*, New Jersey: Prentice Hall Inc.
17.4 Ellis, B. R. and Ji, T. (1996) 'Dynamic testing and numerical modelling of the Cardington steel framed building from construction to completion', *The Structural Engineer*, Vol. 74, No. 11, pp. 186–192.
17.5 Dallard, P., Fitzpatrick, A. J., Flint, A., Le Bourva, S., Low, A., Ridsdill-Smith, R. M. and Willford, M. (2001) 'The London Millennium Footbridge', *The Structural Engineer*, Vol. 79, pp. 17–33.

18 Vibration reduction

18.1 Definitions and concepts

The dynamic vibration absorber (DVA) or tuned mass damper (TMD) is a device which can reduce the amplitude of vibration of a structure through interactive effects. A TMD consists of a mass, connected by means of an elastic element and a damping element to the structure.

- The amplitude of vibration of a structure at resonance can be effectively reduced through slightly increasing or reducing the natural frequency of the structure thus avoiding resonance from an input at a given frequency.
- The amplitude of vibration of a structure can be reduced through base isolation which changes the natural frequencies of the system.
- The amplitude of vibration of a particular mode of a structure can be effectively reduced using a tuned mass damper.

18.2 Theoretical background

Reducing vibration levels induced by different kinds of dynamic loads is a major requirement in the design of some civil engineering structures. The loads include those induced by wind, earthquakes, machines and human activities. For example, some structures prone to wind loads are those which possess a relatively large dimension either in height or length, such as tall buildings and long-span roofs.

Common methods of reducing structural vibration include base isolation, passive energy dissipation and active or semi-active control.

Base isolation: an isolation system is typically placed at the foundation of a structure, which alters the natural frequencies of the structure and avoids resonance from the principal frequency of the excitation force. In earthquake resistance design, an isolation system will deform and absorb some of the earthquake input energy before the energy can be transmitted into the structure.

Passive energy dissipation: dampers installed in a structure can absorb some of the input energy from dynamic loading and/or alter the natural frequencies of the structure, thereby reducing structural vibrations. There are several types of damper used in engineering practice, for example viscoelastic dampers, friction dampers, tuned mass dampers and tuned liquid dampers.

Semi-active and active control: the motion of a structure is controlled or modified by means of a control system through providing external forces which oppose the action of the input.

242 Dynamics

In a similar manner to the previous chapters, limited theoretical background is provided, which serves to explain the demonstration models and practical examples illustrated in this chapter. More detailed information on this topic can be found from [18.1, 18.2 and 18.3].

Altering the structural stiffness, increasing damping of a structure or placing a tuned mass damper is the relatively simple and effective measure to reduce structural vibrations.

18.2.1 Change of dynamic properties of systems

Equation 16.11 has showed that the dynamic response of a single-degree-of-freedom system is the product of the static displacement and the dynamic magnification factor which is defined in equation 16.12. This is shown graphically in Figure 18.1 with damping ratios of $\xi=0.01$, 0.02, 0.03 and 0.05 in the range of frequency ratio between 0.9 and 1.1. Table 18.1 compares the magnification factors of the SDOF system when the frequency ratios are 0.95, 1.0 and 1.05 respectively using equation 16.12. It can be observed from Figure 18.1 and Table 18.1 that:

- Increasing the damping ratio can effectively reduce the vibrations at resonance, simply doubling the damping ratio reduces the peak response by a factor of approximately two.

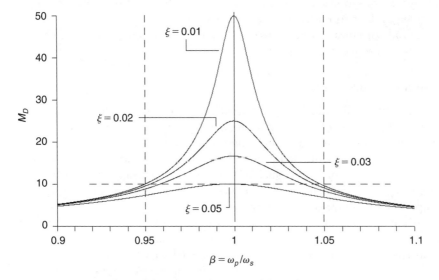

Figure 18.1 Magnification factor as a function of force frequency ratio.

Table 18.1 Magnification factors M_D close to resonance

	$\xi=0.01$	$\xi=0.02$	$\xi=0.03$	$\xi=0.05$
$\beta=0.95$	10.1	9.56	8.85	7.35
$\beta=1.0$	50.0	25.0	16.7	10.0
$\beta=1.05$	9.56	9.03	8.31	6.81

- A slight change of the natural frequency of the SDOF system away from the resonance frequency can effectively reduce the dynamic response. For instance, when the system has a damping ratio of 0.01, the resonance response can be reduced about 80 per cent by increasing or reducing the natural frequency 5 per cent. The same level of vibration reduction requires increasing the damping ratio from 0.01 to 0.05.

Altering the natural frequency of a structure and/or increasing the corresponding damping ratio can reduce vibration at resonance. The concepts behind the two measures are different. Altering the natural frequency of a structure aims to avoid resonance and the structure will experience lower levels of vibration. Increasing damping will increase energy dissipation in a structure, and although the structure will still experience resonance, response reduces through energy dissipation.

Increasing structural stiffness to increase the natural frequency of the structure may not affect the damping ratio or damping mechanism of the structure. The practical case given in section 9.4.4 is a typical example where the floor was stiffened with the profiled external tendons and the fundamental natural frequency of the floor was increased. It was thought that the added tendons would not change the damping mechanism of the floor. It is important to note that structural alterations to increase stiffness may also result in increased mass, thus limiting changes to the natural frequencies.

Reducing structural stiffness reduces the natural frequency of a structure and this approach may be used to avoid resonance. Base isolation systems have been used successfully for the earthquake-resistant design of buildings. An earthquake will shake a building. Placing a base isolation system between the building and the foundation/ground can reduce the level of earthquake forces transmitted to the building. The base isolation system reduces the fundamental natural frequency of the building to a value lower than the predominant frequencies of ground motion to avoid resonance. The first vibration mode of the isolated building is primarily deformation of the isolated system and the building itself appears rigid as illustrated in Figure 18.2. This type of isolation system may absorb a relatively small amount of earthquake energy to suppress any possible resonance at the isolation frequency, but its main function is to change the natural frequency of the building and deform itself. This type of isolation works when the system is linear and elastic, even when it is lightly damped.

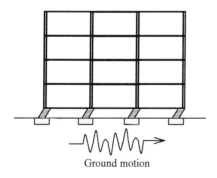

Figure 18.2 Base isolation of a building.

Increasing the damping ratio of a structure by incorporating viscoelastic dampers into the structure may also provide additional stiffness along with a dissipative mechanism. Thus the vibration reduction of the structure due to the added damping materials or dampers could be the combined effect of the increased damping and the increased stiffness. Examples are the London Millennium Footbridge where the viscous dampers were fitted in conjunction with the use of a continuous bracing system as discussed in section 17.4.3 and the test floor panel incorporated with a layer of rubber together with an additional layer of concrete in section 17.4.4.

18.2.2 Tuned mass dampers

The dynamic vibration absorber (DVA) or tuned mass damper (TMD) is a device which can reduce the amplitude of vibration of a system through interactive effects. A TMD consists of a mass, connected by means of an elastic element and a damping element to the system. The principle of TMD reducing vibration of a primary structure is to transfer some structural vibration energy from the structure to the TMD and to split the resonance peak into two less significant resonances. Thus it reduces the vibration of the primary structure while the TMD itself may vibrate significantly.

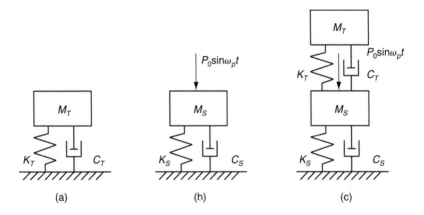

Figure 18.3 A tuned mass damper and a SDOF structure.

Figure 18.3b shows a SDOF system subjected to a harmonic load and its response has been shown in Chapter 16. A tuned mass damper can be placed on the system, forming a two-degree-of-freedom system as shown in Figure 18.3c. The equation of motion of the new system can be described as:

$$\begin{bmatrix} M_S & 0 \\ 0 & M_T \end{bmatrix} \begin{Bmatrix} \ddot{u}_S \\ \ddot{u}_T \end{Bmatrix} + \begin{bmatrix} C_S + C_T & -C_T \\ -C_T & C_T \end{bmatrix} \begin{Bmatrix} \dot{u}_S \\ \dot{u}_T \end{Bmatrix} + \begin{bmatrix} K_S + K_T & -K_T \\ -K_T & K_T \end{bmatrix} \begin{Bmatrix} u_S \\ u_T \end{Bmatrix} = \begin{Bmatrix} P_0 \sin\omega_p t \\ 0 \end{Bmatrix} \quad (18.1)$$

u_S and u_T are the displacements of the SDOF structure system and the TMD respectively; the damping coefficient and stiffness are denoted by C_T and K_T for the TMD and C_S and K_S for the structure. P_0 is the magnitude of the external force applied to the structure system with a frequency ω_p.

Vibration reduction 245

For an appreciation of the efficiency of a TMD in reducing structural vibration, consider a simplified situation where the structural damping is neglected, i.e. $C_S = 0$. Thus the response of the structural system is infinite at resonance without a TMD. Similar to the dynamic magnification factor defined in Chapter 16 for a SDOF system, the dynamic magnification factor for an undamped structural system with a TMD is:

$$R = \sqrt{\frac{(\gamma^2 - \beta^2)^2 + (2\xi_T \gamma \beta)^2}{[(\gamma^2 - \beta^2)^2(1 - \beta^2) - \alpha \gamma^2 \beta^2]^2 + (2\xi_T \gamma \beta)^2(1 - \beta^2 - \alpha \beta^2)^2}} \quad (18.2)$$

where $\beta = \omega_p/\omega_S$ is the forcing frequency ratio; $\gamma = \omega_T/\omega_S$ is the natural frequency ratio; $\omega_S = \sqrt{K_S/M_S}$ is the natural frequency of the structural system; $\omega_T = \sqrt{K_T/M_T}$ is the natural frequency of the TMD system; $\xi_T = C_T/2M_T\omega_T$ is the damping ratio of the TMD system; α is the mass ratio of the TMD to the structure.

It can be seen from equation 18.2 that the magnification factor is a function of the four variables, α, ξ_T, β and γ. Plots of the magnification factor R as a function of the frequency ratio β are given in Figure 18.4a for $\gamma = 1$ and $\alpha = 0.05$ with three ξ_T values (0.0, 0.03 and 0.1), in Figure 18.4b for $\gamma = 1$ and $\xi_T = 0.03$ with three α values (0.01, 0.05 and 1.0) and in Figure 18.4c for $\alpha = 0.05$ and $\xi_T = 0.03$ with three γ values (0.9, 1.0 and 1.1).

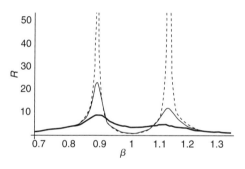

(a) $\gamma = 1$, $\alpha = 0.05$ with $\xi_T = 0.0$ (dashed line), 0.03 (solid line) and 0.1 (dark solid line)

(b) $\gamma = 1$, $\xi_T = 0.03$ with $\alpha = 0.01$ (dashed line), 0.05 (solid line) and 0.1 (dark solid line)

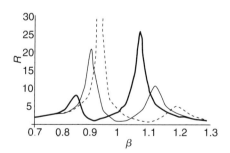

(c) $\alpha = 0.05$, $\xi_T = 0.3$ with $\gamma = 0.9$ (dark solid line), 1.0 (solid line) and 1.1 (dashed line)

Figure 18.4 Dynamic magnification factor as function of β.

246 *Dynamics*

It can be observed conceptually from Figure 18.4 that:

- When $\xi_T=0$ and $\beta=\gamma$, the structural mass is stationary, i.e. the applied load and the force generated by the TMD cancel each other (equation 18.2 and Figure 18.4a).
- The damping in the TMD is important for reducing the vibration at the resonance frequencies of the new two-degrees-of-freedom system (Figure 18.4a).
- The larger the mass ratio, the wider the spread of the two new natural frequencies of the TDOF system, and the smaller the response at $\beta=1$ (Figure 18.4b).
- A TMD is more effective in reducing the vibration of the structure at the natural frequency of the TMD than at that of the structure (Figure 18.4c).
- A TMD can effectively reduce the resonant vibration of a structure with relatively low damping.

It is noted that large movements of a TMD should be considered. In practice several types of TMD have been developed to accommodate different requirements [18.1].

In a TMD a solid (concrete or metal) block often acts as the mass, but in tall buildings a tank filled with water serves the same purpose; this may be considered to be a tuned liquid damper (TLD). Liquids are used to provide not only all the necessary characteristics of the TMD system but also the damping through sloshing action. The mathematical description of TLD response is difficult, but structural implementation is often quite simple.

18.3 Model demonstrations

18.3.1 A tuned mass damper (TMD)

This demonstration *compares free vibrations of a SDOF system and a similar system with a tuned mass damper*, and *shows the effect of the tuned mass damper in free vibration.*

Figure 18.5 shows a single-degree-of-freedom (SDOF) system consisting of a mass and a spring attached to the left hand of a cross arm. The same SDOF system is attached to the right end of the cross arm and a smaller SDOF system is suspended

Figure 18.5 Comparison of vibration with and without a tuned mass damper.

Vibration reduction 247

from it. The smaller SDOF system, which is used as a tuned mass damper, has the same natural frequency as that of the main SDOF system. In other words, the ratio of stiffnesses of the two springs is the same as the ratio of the two masses.

Displace the two main SDOF systems downward vertically by the same amount and then release them simultaneously. It can be seen that the amplitude of vibration of the SDOF system on the left is greater than that of the one on the right. The tuned mass damper vibrates more than the mass to which it is attached. The vibration of the tuned mass damper applies a harmonic force to the SDOF system, which suppresses the movements of the main mass.

The effect of a tuned mass damper is most significant if a harmonic load is applied to the system with a frequency close to the natural frequency of the system.

18.3.2 A tuned liquid damper (TLD)

This demonstration *compares free vibration of two SDOF systems, one with and one without water*, and *shows the effect of a turned liquid damper in reducing free vibration*.

A tuned liquid damper (TLD) is basically a tank of liquid that can be tuned to slosh at the same frequency as the structure to which it is attached.

Two circular tanks are attached to the top of two identical flexible frames as shown in Figure 18.6. The tank on the right is filled with some coloured water. The effect of water sloshing on the vibration of the supporting structure can be demonstrated as follows:

1 Displace the tops of the two frames laterally by similar amounts.
2 Release the two frames simultaneously and observe free vibrations of the two frames. The amplitude of vibration of the frame on the right decays much more quickly than that of the frame on the left.

Figure 18.6 Comparison of vibration with and without the effect of liquid.

248 *Dynamics*

The difference is due to the water sloshing in the tank and dissipating the energy of the system.

18.3.3 Vibration isolation

This demonstration *compares forced vibrations of two glasses containing similar amounts of water, one with and one without a plastic foam support*, and *shows the effect of base isolation*.

A medical shaker can be used as a shaking table to generate harmonic base movements in three perpendicular directions. A glass is fixed directly to a wooden board while a similar glass is glued to a layer of plastic foam which is mounted on the wooden board, as shown in Figure 18.7.

Fill the two glasses with similar amounts of water. When the shaker moves at a preset frequency, it can be seen that the water in the glass on the plastic foam moves less than that in the other glass. The difference is due to the effect of the plastic foam, which isolates the base motion and also produces a natural frequency of the glass–water–foam system which is lower than that of the glass–water system and is thus farther away from the vibration frequency.

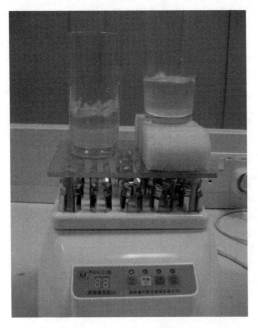

Figure 18.7 Altering structural frequency using base isolation.

18.4 Practical examples

18.4.1 Tyres used for vibration isolation

Figure 18.8 shows two tyres placed between the ground and a generator in a rural area of a developing country. The presence of the tyres reduced the natural frequency of the generator and moved it away from the operating frequency. Although the operators of the generator may not have known much about vibration theory,

Figure 18.8 Tyres used for base isolation (courtesy of Professor B. Zhuang, Zhejiang University, China).

they knew from experience that the presence of the tyres could reduce the level of vibration.

18.4.2 The London Eye

The British Airways London Eye is the largest observation wheel in the world (Figure 18.9a). It was built by the River Thames in the heart of London, just opposite to the Palace of Westminster. The wheel has a height of 135m and carries 32 capsules which can hold up to 800 people.

In order to reduce vibrations in the direction perpendicular to the plane of the wheel due to wind loads, 64 tuned mass dampers were installed in steel tubes which are uniformly distributed around the wheel. One of these tubes is indicated in Figure 18.9b. The relation between the tuned mass damper and the tube is illustrated in Figure 18.10. A mass and a spring in the tube were designed with a natural frequency close to the natural frequency of the wheel in the direction perpendicular to the plane of the wheel. Plenty of room is available in the tube to allow large movements of the tuned mass damper. No excessive vibration of the wheel has been observed in this direction since it was erected [18.4].

18.4.3 The London Millennium Footbridge

A total of 26 pairs of vertical tuned mass dampers were installed over the three spans of the London Millennium Footbridge [18.5]. These comprise masses of

(a) Front view

(b) A tuned mass damper placed inside of a tube

Figure 18.9 The London Eye.

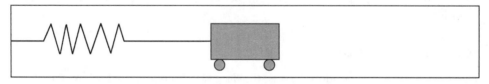

Figure 18.10 A tuned mass damper in a tube of the London Eye.

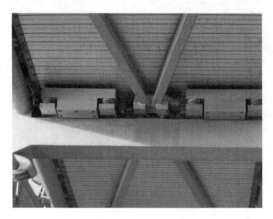

Figure 18.11 Two tuned mass dampers under the deck of the London Millennium Footbridge.

between one and three tonnes supported on compression springs. The tuned mass dampers are situated on the top of the transverse arms beneath the deck. One pair of the tuned mass dampers is shown in Figure 18.11.

These tuned mass dampers would become effective if the footbridge experienced relatively large vertical vibrations, though such vibrations were not the primary source of the well-publicised problems when the bridge first opened.

References

18.1 Soong, T. T. and Dargush, G. F. (1997) *Passive Dissipation Systems in Structural Vibration*, New York: John Wiley & Sons.
18.2 Soong, T. T. (1990) *Active Structural Control: Theory and Practice*, Harlow: Longman Scientific & Technical.
18.3 Korenev, B. G. and Reznikov, L. M. (1993) *Dynamic Vibration Absorbers*, New York: John Wiley & Sons.
18.4 Rattenbury, K. (2006) *The Essential Eye*, London: HarperCollins Publishers.
18.5 Dallard, P., Fitzpatrick, A. J., Flint, A., Le Bourva, S., Low, A., Ridsdill-Smith, R. M. and Willford, M. (2001) 'The London Millennium Footbridge', *The Structural Engineer*, Vol. 79, pp. 17–33.

19 Human body models in structural vibration

19.1 Concepts

- A stationary person, i.e. one who is sitting or standing, acts as a mass-spring-damper rather than as an inert mass in both vertical and lateral structural vibration.
- A walking or jumping person acts solely as dynamic loading on structures and thereby induces structural vibration.
- A bouncing person acts as both dynamic loading and a mass-spring-damper on structures in vertical structural vibration.

19.2 Theoretical background

19.2.1 Introduction

How people interact with their environment is a topical issue and one of increasing importance. One form of physical interaction which is understood poorly, even by professionals, is concerned with human response to structural vibration. This is important, for example, when determining how dance floors, footbridges and grandstands respond to moving crowds and for determining how stationary people are affected by vibration in their working environment. Human–structure interaction provides a new topic that *describes the independent human system and structural system working as a whole and studies the structural vibration where people are involved and human body response to structural movements.*

When a structure is built on soft soil, the interaction between the soil and the structure may be considered; when a structure is in water, such as an offshore platform, the interaction between the structure and the surrounding fluid may be considered. Similarly when a structure is loaded with people, the interaction between people and structure may need to be considered. An interesting question is why has this not been considered earlier? There are two reasons:

1. The human body is traditionally considered as an inert mass in structural vibration. For example, Figure 19.1 is a question taken from a well-known textbook on Engineering Mechanics [19.1] where the girl is modelled as an inert mass in the calculation of the natural frequency of the human–beam system.
2. The human mass is small in comparison with the masses of many structures and in this situation its effect is negligible. Thus there have been no requirements from practice for considering such effects as human–structure interaction.

Human body models in structural vibration 253

Figure 19.1 A girl standing on a beam [19.1] (permission of John Wiley & Sons Inc.).

Dynamic measurements were taken on the North Stand at Twickenham when it was empty and when it was full of spectators (section 19.4.1). The observations on the stand suggested a new concept that *the stationary human whole-body acts as a mass-spring-damper rather than an inert mass in structural vibrations*. The phenomenon was reproduced in the laboratory (section 19.3.1). It was confirmed that *the stationary human whole-body did not act as an inert mass but appeared to act as a mass-spring-damper system in structural vibrations* [19.2].

Today, many structures are lighter and have longer spans than earlier similar types of construction and as a consequence the effect of human bodies has become more important. The newly emerged problems are the human-induced vibrations of grandstands and footbridges, where crowds of people are involved, and the human perception of structural vibration induced by human actions.

The study of human–structure interaction is concerned with both structural dynamics and body biomechanics [19.3, 19.4]. The former belongs to engineering while the latter is part of science. The following diagram describes the study of structural dynamics:

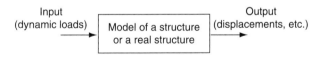

Figure 19.2 The basic studies in structural dynamics.

The structure may range from a simple beam to a complex building, or from a car to an aeroplane. The relationships between input, output and the model of a structure can normally be described by governing equations and the solution of the equations is the output. In the diagram the input, output and the structure can be quantified or at least quantified statistically.

If a similar diagram to Figure 19.2 is required to describe the basic studies in biomechanics of the human body, it may be represented as follows:

Figure 19.3 The basic study in body biomechanics.

The objective of the study of human response to vibration is to establish relationships between causes and effects [19.5]. However, there are no governing equations available to describe the relationships between causes, people and effects. This might be because one cause may generate a range of effects and/or different causes induce the same effect. In addition, the effects, relating to comfort, interference and perception of vibration, may be descriptive and difficult to quantify.

As human–structure interaction is a new topic and the problems in practice have only recently emerged, there is only limited information available on the topic. Yet understanding the human whole-body models in structural vibration and the dynamic properties of the whole-body subject to low amplitude vibration are critical in the development of this new topic. The concept of human–structure interaction has been partly considered in a design code and a design guidance [19.6, 19.7].

In the future it is probable that structures will have longer spans and be lighter, and the human expectation of quality of life and working environment will be greater. Therefore, engineers will need an improved understanding of human–structure interaction to tackle these problems where structural serviceability and/or human comfort are concerned.

19.2.2 Identification of human body models in structural vibration

Human whole-body models in structural vibration can be qualitatively identified through experimental methods by placing the human body on a vibrating structure.

A structure is considered as a single-degree-of-freedom (SDOF) system as shown in Figure 19.4b, which has a natural frequency of $\omega_S = \sqrt{K_S/M_S}$ where K_S and M_S are the stiffness and mass of the SDOF system. If a human body acts as an inert mass M_H on the SDOF structure system (Figure 19.4c), the natural frequency of the human–structure system becomes:

$$\omega_{HS} = \sqrt{\frac{K_S}{M_S + M_H}} < \sqrt{\frac{K_S}{M_S}} = \omega_S \tag{19.1}$$

Therefore the natural frequency of the human–structure system would be less than that of the structure system.

If a human whole-body is considered as a mass-spring-damper system with coefficients M_{H1}, K_{H1} and C_{H1} which are consistent with the vibration of the first mode of the body as shown in Figure 19.5a, its natural frequency is:

$$\omega_H = \sqrt{\frac{K_{H1}}{M_{H1}}} \tag{19.2}$$

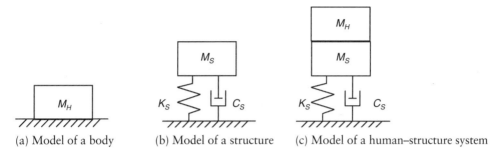

Figure 19.4 A human–structure model when the body acts as an inert mass.

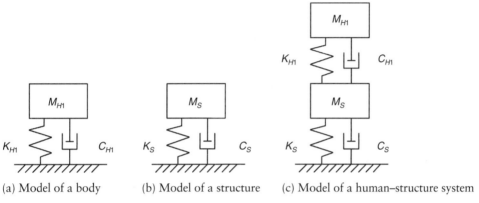

Figure 19.5 A human–structure model when the body acts as a mass-spring-damper.

Placing the SDOF human body model (Figure 19.5a) on to the SDOF structure system (Figure 19.5b) forms a two-degrees-of-freedom (TDOF) system (Figure 19.5c). The equation of motion of the human–structure system is the same as that for a SDOF structure system with a tuned mass damper attached.

The natural frequencies of the combined human–structure system (Figure 19.5c) can be obtained by solving the corresponding eigenvalue problem of the following equations of motion:

$$\begin{bmatrix} M_S & 0 \\ 0 & M_{H1} \end{bmatrix} \begin{Bmatrix} \ddot{u}_S \\ \ddot{u}_H \end{Bmatrix} + \begin{bmatrix} C_S+C_{H1} & -C_{H1} \\ -C_{H1} & C_{H1} \end{bmatrix} \begin{Bmatrix} \dot{u}_S \\ \dot{u}_H \end{Bmatrix} + \begin{bmatrix} K_S+K_{H1} & -K_{H1} \\ -K_{H1} & K_{H1} \end{bmatrix} \begin{Bmatrix} u_S \\ u_H \end{Bmatrix} = \begin{Bmatrix} 0 \\ 0 \end{Bmatrix} \quad (19.3)$$

By neglecting the damping terms a harmonic solution that satisfies the above equation can be determined as:

$$\begin{Bmatrix} u_S \\ u_H \end{Bmatrix} = \begin{Bmatrix} A_S \\ A_H \end{Bmatrix} \sin(\omega t + \phi) \quad (19.4)$$

The symbols A_S and A_H in equation 19.4 represent the amplitudes of the vibration of the system. Substituting equation 19.4 into equation 19.3 gives the following equations:

$$\begin{bmatrix} K_S+K_{H1}-\omega^2 M_S & -K_{H1} \\ -K_{H1} & K_{H1}-\omega^2 M_{H1} \end{bmatrix} \begin{Bmatrix} A_S \\ A_H \end{Bmatrix} = \begin{Bmatrix} 0 \\ 0 \end{Bmatrix} \quad (19.5)$$

256 Dynamics

For convenience the following terms are defined:

$$\alpha = \frac{M_{H1}}{M_S} \quad \gamma = \frac{\omega_H}{\omega_S}$$

where α and γ are called the modal mass ratio and the frequency ratio of the SDOF human body system to the SDOF structure system, and have positive values. Solving equation 19.5 gives the natural frequencies of the TDOF system ω_1 and ω_2 represented using ω_S, ω_H and α:

$$\omega_1^2 = \frac{1}{2}\left[\omega_S^2 + \alpha\omega_H^2 + \omega_H^2 - \sqrt{(\omega_S^2 + \alpha\omega_H^2 + \omega_H^2)^2 - 4\omega_S^2\omega_H^2}\right] \quad (19.6a)$$

$$\omega_2^2 = \frac{1}{2}\left[\omega_S^2 + \alpha\omega_H^2 + \omega_H^2 + \sqrt{(\omega_S^2 + \alpha\omega_H^2 + \omega_H^2)^2 - 4\omega_S^2\omega_H^2}\right] \quad (19.6b)$$

Thus frequency relationships between the human–structure system (ω_1, ω_2) and the independent human and structure systems (ω_S, ω_H) can be found as follows:

Relationship 1:

$$\omega_1^2 + \omega_2^2 = \omega_S^2 + (1+\alpha)\omega_H^2 > \omega_S^2 + \omega_H^2 \quad (19.7)$$

This relationship indicates that *the sum of square of the natural frequencies of the combined human–structure system is larger than that of the corresponding human and structure systems*. Equation 19.7 is obtained by adding equation 19.6a to equation 19.6b.

Relationship 2:

$$\omega_1\omega_2 = \omega_S\omega_H \quad (19.8)$$

This relationship indicates that *the product of natural frequencies of the human–structure system equals that of the corresponding human and structure systems*. Equation 19.8 can be derived by multiplying equation 19.6a and equation 19.6b.

Relationship 3:

$$\omega_1 < (\omega_S, \omega_H) < \omega_2 \quad (19.9)$$

This relationship indicates that *the natural frequencies of the human and structure systems are always between those of the human–structure system*. Equation 19.9 can be determined as follows:

$$\frac{\omega_2}{\omega_S} = \sqrt{\frac{1.0 + \alpha\gamma^2 + \gamma^2 + \sqrt{(1.0 + \alpha\gamma^2 + \gamma^2)^2 - 4\gamma^2}}{2.0}}$$

$$> \sqrt{\frac{1.0+\alpha\gamma^2+\gamma^2+\sqrt{(1.0+\gamma^2)^2-4\gamma^2}}{2.0}}$$

$$= \sqrt{\frac{1.0+\alpha\gamma^2+\gamma^2+\sqrt{(1.0-\gamma^2)^2}}{2.0}} = \sqrt{1+\alpha\gamma^2/2} > 1.0 \tag{19.10}$$

Substituting equation 19.10 into relationship 2 (equation 19.8) leads to relationship 3.

The three relationships are valid without any limitation on the values of the natural frequencies of the human and structure systems and valid for any system that can be represented as a TDOF system.

Equation 19.1 and equation 19.9 provide the qualitative relationships to identify human body models in structural vibration through experiments.

19.3 Demonstration tests

19.3.1 *The body model of a standing person in the vertical direction*

The tests demonstrate that *a standing person acts as a mass-spring-damper rather than an inert mass in vertical structural vibration while a walking or jumping person acts solely as loading on structures.*

Figure 19.6a shows a simply supported reinforced concrete beam with an accelerometer placed at the centre of the beam. Striking the middle of the beam vertically with a rubber hammer caused vertical vibrations of the beam. The acceleration–time history recorded from the beam and the frequency spectrum abstracted from the record are shown in Figures 19.7a and 19.7b. It can be observed that the simply supported beam has a vertical natural frequency of 18.7 Hz and the system has a very small damping ratio shown by the decay of the free vibrations which lasts more than eight seconds.

A person then stood on the centre of the beam as shown in Figure 19.6b and a human–structure system was created. A rubber hammer was again used to induce vibrations. Figures 19.7c and 19.7d show the acceleration–time history and frequency spectrum of the beam with the person. Comparing the two sets of measurements in Figure 19.7 shows that:

- The measured vertical natural frequency of the beam with the person was 20.0 Hz which is larger than that of the beam alone (Figures 19.7b and 19.7d). This observation coincides with relationship 3, i.e. $\omega_S < \omega_2$.
- The beam with the person possesses a much larger damping ratio than the beam alone as the free vibration of the beam with the person decays very quickly (Figure 19.7c). This is also evident from Figure 19.7d as the peak in the spectrum of the beam with the person has a much wider spread than that of the beam alone (Figure 19.7a).

Further tests were conducted to identify the effects qualitatively of human bodies in structural vibration, including tests on the beam with a dead weight and the beam

(a) An empty beam

(b) A person standing on the beam

Figure 19.6 Test set-up for identifying human body models in vertical structural vibration.

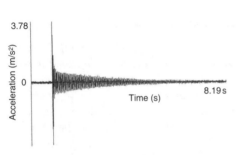
(a) Free vibration of the beam alone

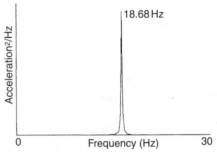
(b) Response spectrum of the beam alone

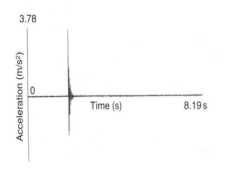
(c) Free vibration of a human–beam system

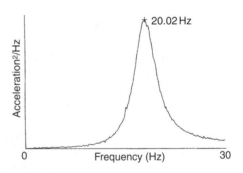
(d) Response spectrum of the human–beam system

Figure 19.7 Measurements of the identification tests in vertical directions [19.2].

occupied by a person who moved, both jumping and walking. The measured frequencies for these cases are listed in Table 19.1, which shows:

- The dead weights, which were placed centrally on the beam for two tests, reduced the natural frequency as expected. This can be accurately predicted using equation 19.1.
- The test with the person standing on the beam showed an increase in the measured natural frequency. This observation cannot be explained using the inert mass model or equation 19.1. Thus it is clear that the standing human body does not act as an inert mass in structural vibration.
- The measured frequency for the vibrations when the person sat on a stool on the beam was also higher than that of the beam alone.
- Significant damping contributions from the human whole-body were observed for both standing and sitting positions, as can be appreciated from Figure 19.7 for the standing person.
- Jumping and walking also provided interesting results in that they did not affect either natural frequency or damping. The unchanged system characteristics would appear to be because the moving human body is not vibrating with the beam.

Table 19.1 Natural frequencies observed on the beam [19.2]

Description of experiments	Measured natural frequency (Hz)
Bare beam (Figure 19.6a)	18.7
Beam plus a mass of 45.4 kg (100 lb)	15.8
Beam plus a mass of 90.8 kg (200 lb)	13.9
Beam with T. Ji standing (Figure 19.6b)	20.0
Beam with T. Ji sitting on a high stool	19.0
Beam with T. Ji jumping on spot	18.7
Beam with T. Ji walking on spot	18.7

Two concepts can be identified from the above tests:

- *A stationary person, e.g. sitting or standing, acts as a mass-spring-damper rather than as an inert mass in structural vibration.*
- *A walking or jumping person acts solely as loading on structures.*

19.3.2 *The body model of a standing person in the lateral direction*

The tests demonstrate that *a standing person acts as a mass-spring-damper rather than an inert mass in lateral structural vibration* [19.8].

Figure 19.8a shows a single-degree-of-freedom rig for both vertical and horizontal directions used for identification tests of human body models in structural vibration. The test rig consists of two circular top plates bolted together, three identical springs supporting the plates and a thick base plate. The test procedure is simple and is the same as that conducted in section 19.3.1. The free vibration test of the test rig alone was first conducted using a rubber hammer to generate an impact on the rig in

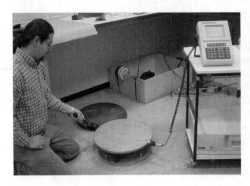

(a) A SDOF test rig   (b) A person standing on the test rig

Figure 19.8 Test set-up for identifying human body models in lateral structural vibration.

the lateral direction. Then a person stood on the test rig and an impact was applied in the lateral direction parallel to the shoulder of the test person as shown in Figure 19.8b. Figure 19.9 shows the displacement-time histories and the corresponding spectra of the test rig alone and the human-occupied test rig in the lateral directions. It can be noted from Figure 19.9 that:

- The standing body contributes significant damping to the test rig in the lateral direction (Figures 19.9a and 19.9c).
- There is one single resonance frequency recorded on the test rig alone (Figure 19.9b) but two resonance frequencies are observed from the human–structure system in the lateral direction (Figure 19.9d).
- The single resonance frequency of the test rig alone is between the two resonance frequencies of the human-occupied test rig (relationship 3 in section 19.2.2).
- Human whole-body damping in the lateral directions is large but less than that in the vertical direction as shown by the vibration time history of the human–structure system in the lateral directions which is longer than that in the vertical directions (Figure 19.7c), although the test structures are different.

The experimental results of the identification tests conducted in the lateral directions clearly indicate that *a standing human body acts in a similar manner to a mass-spring-damper rather than an inert mass in lateral structural vibration.*

Further identification tests have been conducted on the same test rig with a bouncing person who maintains contact with the structure. It is observed that:

- *A bouncing person acts as both loading and a mass-spring-damper on structures in vertical structural vibration.*
- *The interaction between a bouncing person and the test rig is less significant than that between a standing person and the test rig.*

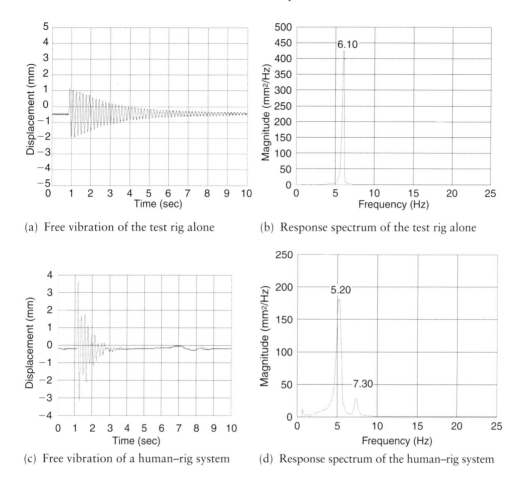

Figure 19.9 Measurements of the identification tests in lateral directions.

19.4 Practical examples

19.4.1 The effect of stationary spectators on a grandstand

Measurements were taken to determine the dynamic behaviour of the North Stand (Figure 19.10) at the Rugby Football Union ground at Twickenham. The grandstand has three tiers and two of them are cantilevered. Dynamic tests were performed on the roof and cantilevered tiers of the stand, with further measurements of the response of the middle-cantilevered tier to dynamic loads induced by spectators during a rugby match [19.2].

Spectra for the empty and full grandstand are given in Figure 19.11. Figure 19.11a shows a clearly defined fundamental mode of vibration for the empty structure. Instead of the expected reduction in natural frequency of the stand as the crowd assembled, the presence of the spectators appeared to result in the single natural frequency changing into two natural frequencies (Figure 19.11b). Figure

262 Dynamics

19.11 shows that the dynamic characteristics of the grandstand changed significantly when a crowd was involved and that the structure and the crowd interacted. This pattern was also noted in two other locations of the stand where measurements were taken.

Comparing the spectra for the empty stand and fully occupied stand, three significant phenomena are apparent:

- *an additional frequency was observed in the occupied stand;*
- *the natural frequency of the empty stand is between the two natural frequencies of the occupied stand;*
- *the damping increases significantly when people were presented.*

Considering human bodies simply as masses cannot explain the above observations. The observations suggest that the crowd acted as a mass-spring-damper rather than just as a mass. When the crowd is modelled as a single-degree-of-freedom system, the structure and the crowd form a two-degrees-of-freedom system. Based on this model, the above observations can be explained. These observations complement the laboratory tests described in section 19.3.

Figure 19.10 The North Stand, Twickenham.

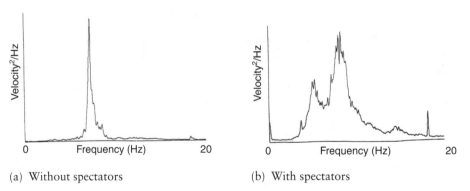

(a) Without spectators (b) With spectators

Figure 19.11 Response spectra of the North Stand, Twickenham.

19.4.2 Calculation of the natural frequencies of a grandstand

It is common for a grandstand, either permanent or temporary, to be full of spectators during a sports event or a pop concert as shown in Figure 19.12. In this situation, the human mass can be of the same order as the mass of the structure. When calculating the natural frequencies of a grandstand for design purposes, it is necessary to consider how the human mass should be represented. It is recommended in BS6399: Part 1: Loading for Buildings [19.6] and the Interim Guidance [19.7] that empty structures should be used for calculating their natural frequencies, i.e. the human mass should not be included. This is because the worst situation for design consideration is when people move rather than when people are stationary. Therefore engineers will not underestimate the natural frequencies of grandstands through adding the mass of a crowd to the structural mass.

Figure 19.12 A grandstand full of spectators.

19.4.3 Dynamic response of a structure used at pop concerts

During a pop concert, it is rare that everyone moves in the same way following the music. Some people will move enthusiastically, jumping or bouncing, some will sway and some will remain stationary, either sitting or standing. Those who are stationary will provide significant damping to the structure as observed from the previous demonstrations and thus will alter the dynamic characteristics of the structure and effectively damp the level of vibration induced by the movements of the others. This in part explains why the predicted structural vibrations induced by human movements, where stationary people are not considered, are often much larger than vibrations measured on site.

The effect of stationary people in structural vibration has been considered in the new version of the Interim Design Guidance [19.7] for predicting the response of grandstands used for pop concerts.

19.4.4 Indirect measurement of the fundamental natural frequency of a standing person

If a stationary person should be modelled as a single-degree-of-freedom (SDOF) system in structural vibration, what are the natural frequency, damping ratio and the modal mass of the human SDOF system?

264 *Dynamics*

The natural frequency of a human body cannot be obtained directly using traditional methods and tools of structural dynamics, such as sensors (accelerometers) which cannot be conveniently mounted on a human body. However, a method has been developed to estimate the natural frequency of a standing person by experiment without touching the person whilst still using the methods of structural dynamics.

A simple formula can be derived based on the three frequency relationships derived in section 19.2.2 using the measurements of the natural frequency of the empty structure and the resonance frequency/frequencies of the human–structure system as demonstrated in section 19.3. For example, the demonstration given in section 19.3.2 shows the natural frequency of the test rig alone of 6.10 Hz and the two resonance frequencies of the human–rig system of 5.20 Hz and 7.30 Hz respectively in Figure 19.9. Using equation 19.8 gives an estimated natural frequency of the standing person of 6.22 Hz in the lateral direction. It should be noted that this is an approximation because:

- The measurements in Figure 19.9d are the resonance frequencies which include the effect of human body damping while equation 19.8 does not consider any effect of damping.
- The simple mass-spring-damper model is good enough for identifying the models of a human body in structural vibration qualitatively. However, this body model is developed on fixed ground rather than on a vibrating structure.

The indirect measurement method needs to be improved; nevertheless it uses the methods of structural dynamics to study the biomechanics properties of a human body, which is fundamentally different from the methods of body biomechanics where shaking tables are used [19.9].

19.4.5 *Indirect measurement of the fundamental natural frequency of a chicken*

Six hundred million chickens are consumed each year in the UK. It has been observed sometimes during transportation that healthy chickens become ill and are unable to stand. Possible causes for this have been found to arise from the effect of resonance associated with the transportation.

In order to prevent the resonance in which a natural frequency of the truck matches the body natural frequency of chickens, the natural frequency of a typical chicken needs to be identified. When studying body biomechanics to determine the characteristics of a human body, a subject is asked to sit or to stand on a shaking table for a short period of vibration. However, this technique cannot be applied to studying the natural frequency of a chicken, as the chicken will fly off when the shaking table moves.

The method developed for indirectly measuring human body natural frequencies could be used to obtain the natural frequency of a chicken. Figure 19.13 shows a chicken perched on a wooden beam. A slight impact on the beam can generate vibration of the beam and the chicken, which will not cause anxiety in the chicken. When the frequencies of the bare beam and the beam with the chicken are measured, the natural frequency of the chicken can be estimated from the two measurements and a simple equation.

Human body models in structural vibration 265

Figure 19.13 Indirect measurement of the natural frequency of a chicken (courtesy of Dr J. Randall).

References

19.1 Meriam, J. L. and Kraige, L. G. (1998) *Engineering Mechanics, Vol. 2: Dynamics*, Fourth Edition, New York: John Wiley & Sons.
19.2 Ellis, B. R. and Ji, T. (1997) 'Human–structure interaction in vertical vibrations', *Structures and Buildings, the Proceedings of Civil Engineers*, Vol. 122, No. 1, pp. 1–9.
19.3 Ji, T. (2000) 'On the combination of structural dynamics and biodynamics methods in the study of human–structure interaction', The 35th United Kingdom Group Meeting on Human Response to Vibration, Southampton, England, 13–15 September.
19.4 Ji, T. (2003) 'Understanding the interactions between people and structures', *The Structural Engineer*, Vol. 81, No. 14, pp. 12–13.
19.5 Griffin, M. J. (1990) *Handbook of Human Vibration*, London: Academic Press Limited.
19.6 BSI (1996) BS 6399: Part 1: Loading for Buildings, London.
19.7 Institution of Structural Engineers (2001) *Dynamic Performance Requirements for Permanent Grandstands Subject to Crowd Action – Interim Guidance on Assessment and Design*, London.
19.8 Duarte, E. and Ji, T. (2006) 'Measurement of human–structure interaction in vertical and lateral directions: a standing body', ISMA International Conference on Noise and Vibration Engineering, Leuven, Belgium.
19.9 Matsumoto, Y. and Griffin, M. (2003) 'Mathematical models for the apparent masses of standing subjects exposed to vertical whole-body', *Journal of Sound and Vibration*, Vol. 260, No. 3, pp. 431–451.

Index

amplitude of motion
   forced harmonic vibration 212, 217–19
   free vibration 187–9
angle of twist 50

base isolation 243, 248
beams
   overhanging 40
   simply supported 39
bending 38–46
   assumption 42
   failure 44
   profiles of girders 43
   staple 45
bending moment 38–41
   diagrams 38
   relationship to load and shear force 38
boundary conditions 5, 71–4, 117, 120–1
bracing systems 82–94
   scaffolding structures 93–96
   tall buildings 92
buckling
   box girder 125
   boundary condition 117, 120–1
   bracing member 124
   column 113
   empty can 123
   shapes 119

cable structure 137
calculation of section properties 13–24, 28–35
   centroid 14–17
   I-section 31
   parallel axes theorem 28
   rectangular section 31
   V-section 32
centre of gravity 13
centre of mass 13–27
   body in a horizontal plane 19–21
   body in a vertical plane 21–2
   to centroid of a body 19
   crane 24

display unit 26
Eiffel Tower 25
Kio Towers 26–7
to motion 24
piece of cardboard of arbitrary shape 18–19
to stability 17, 22–4
centroid 13–18
columns 113–17, 119–21
conditions of equilibrium 4
conservation 159–69
   of energy 160–5, 167–9
   of momentum 165
conservative system 160
critical buckling load 116, 118
critical damping 186–91

damped systems 187–91, 201, 211–20
   critically damped 187
   overcritically damped 187–9, 201
   undercritically damped 189–91, 211–19
damping effect 187–91, 201, 211–19
   free vibration 187–91, 201
   response to harmonic excitation 211–19
damping ratio 186, 220
deflection 67–76, 79, 92
   column support 74
   effect of boundary condition 72
   effect of span 71
   metal prop 75–6
   prop root 75
deflection of beam 67–74
   differential equation of bending 67
   sign convention 68
direct force paths 77–95
   braced frames 82–95
   concepts 77
   criteria for bracing 82
   demonstration model 92
   experiment 90
   temporary grandstand 87
dynamic magnification factor 213–14, 218–19

effective length 116
elastic modulus 67, 116
energy 159–64
   gravitational 159
   kinetic 160, 163–4
   strain 159
energy conservation 159–69
   collision balls 165–6
   dropping a series of balls 167–8
   moving wheel 163–4
   rollercoaster 168
   torch without energy 169
energy method 79, 81, 97–104, 159–69
equation of beam deflection 67
equation of motion
   free vibration 186
   SDOF systems subjected to harmonic ground motion 217
   SDOF systems subjected to a harmonic load 211
equilibrium 3–12
   action and reaction force 5
   barrier 8
   dust train 11
   footbridge 9
   kitchen scale 10
   magnetic floating model 7
   magnetic floating train 11
   plate-bottle system 7
   stable and unstable 6
   stage performance 10
equivalent horizontal load 141, 145, 149
experimental testing
   building 206–7, 227–8
   chicken–structure system 264–5
   floor 238–9
   forced harmonic vibration 227–8
   free vibration 206–7
   grandstand 261–2
   human–structure system 257–64
   resonance 227–8
   stack 207–8

fixed end moment 73
forces
   bending 38–42
   internal 79–87, 96–104
   normal 61–6, 127–32
   shear 38, 47–60
   torsion 49–53, 56–7, 59
forced vibration 210–20
   damped 211–20
   frequency ratio for 212–14, 217–20
   resonance frequency of 219–20
Foucault pendulum 182–4
frame 140–53
free body diagram 5, 39

free vibration 185–209
   damped 186–91
   decay 186–91, 200–1
   music box 204–6
   equation for 186
frequency
   natural and circular 186
   resonant 219–20
frequency ratio 212–14, 218–19

generalised
   mass 191–5
   stiffness 191–5
generalised SDOF systems 191
gravitational force 4
gravity 13

harmonic vibration (forced) 211–17
horizontal movement of frames 139–56
   building floor 153
   grandstand 153
   model demonstrations 151–3
   rail bridges 156
human–structure interaction 252–65

identification
   human body model 257–61
   chicken 264–5
integration method 69–70
internal forces
   in beams 81, 97–105
   in frames 140–9
   in trusses 79–87

joints
   pinned 79–81
   rigid 81

Lagrange equation 163–4
lateral-torsional buckling of beams 116–19, 122–3, 125–6
   demonstrations 122–3
   prevention 125–6
load factor 141–9
load, types of
   axial 127–32
   bending moment 3
   concentrated 5
   distributed 39
   externally applied 3, 39
   internal forces 39, 96
   load, shear force and bending moment relationships 38
   normal force 61
   shear 48–9
   torque 50–2

268   Index

magnification factor   213–15, 218–19
mass   186, 211
mass-spring-damper system   186, 244–7, 249–50
mode shape   192, 202, 203
mode superposition   196
modulus of elasticity   67, 116
momentum   159
motion   186, 211

natural frequency   185–6, 196–9, 203–4
   to displacement   196–8
   to tension force   198–9, 203–4
natural period   185
neutral plane, neutral axis   32, 38, 42
normal force   61–6

parallel-axis theorem   29, 32
pendulum   170–84, 199
phase angle   214–17
pin-jointed structure   79–87
pinned support   4
prestress   127–38
   centrally applied   128
   demonstration   133–4
   eccentrically applied   128–30
   externally applied   130–2
   practical examples   134–7

Raleigh Arena   107–8
relationship
   between natural frequency and displacement   196–8
   between natural frequency and tension force   198–9, 203–4, 208–9
resonance
   avoidance   225–6
   effect   221–2
   footbridge   223–5
   SDOF system   221
   washbowl   228–9
resonance frequency   219–20
resonance testing   227–8

second moment of area   28–35
shape function   193–5
shear force   38–41
shear modulus   50, 118
shear stress   47–60
   in bending   47–9
   effect   53–60
   in torsion   47, 49–52
sign convention   68
simple harmonic motion   211
simply supported beam   5, 39–42, 192–5
single-degree-of-freedom system   186–91, 211–23

smaller internal forces   96–112
   cable-stayed bridge   111
   floor structure   111–12
   pair of rubber rings   105–6
   Raleigh Arena   106–7
   Zhejiang Dragon Sports Centre   108–9
spider's web   135–7
stable and unstable equilibria   6
stiffness   79–81, 191–6
   at critical point   79
   at point   79
   lateral   81
   modal   191–6
strain energy
   bending   79, 81, 163
   tension and compression   79
stress
   normal stress due to bending   30
   normal stress due to tension and compression   61–6
   shear stress due to bending and torsion   47–51
stress distribution   62–6
   balloons on nails   62–3
   flat shoes vs high-heel shoes   64–5
   the Leaning Tower of Pisa   65–6
   uniform and non-uniform   63
structure factor   141–9
strut   115–21
superposition   70, 196
support
   fixed   4
   pinned   4
   roller   4
suspended system   170–84
   effect of added mass   178–80
   floor   182–3
   generalised   172–6
   inclined suspended wood bridge   182
   natural frequency   171–80
   rotational   176–80
   static behaviour   180–1
   translational   176–80

tension   127–37
tension structure   137
torque   49–53
torsion   49–53
   circular section   50–1
   non-circular section   50–3
transmissibility   218
trusses   79
tuned-liquid-damper   247
tuned-mass-damper   244–7, 249–51
two-degrees-of-freedom system   244–7, 254–61

unit load method 79–87

vibration
  amplitude 212–14, 217–19, 242
  damped 186, 211
  forced 211–29
  free 185–209
  reduction of 241–50
vibration absorber 244–8
vibration isolation 248–9

vibration reduction 241–50
  London Eye 249–50
  London Millennium Footbridge 249
viscous damping 186, 212, 234–5

warping 56, 57
work 79, 81

Young's modulus 67, 116